Current State and Future Perspective in Human-Robot Interaction

Edited by Constantin Voloşencu

Published in London, United Kingdom

Current State and Future Perspective in Human-Robot Interaction
http://dx.doi.org/10.5772/intechopen.1003383
Edited by Constantin Voloşencu

Contributors
Adrian Prados, Alberto Mendez, Alex Alspach, Alicia Mora, Caroline Hands, David Crandall, Gary Leander, Gerardo Iovane, Gloria Álvarez, Gonzalo A. García, Gonzalo Espinoza, Guillermo Pérez, Iana Fominska, Ibidun Christiana Obagbuwa, Ioannis Angelakis, José Gabriel Amores, Katherine M. Tsui, Kelvin Cheng Kian Tan, Krishnasamy Tamilselvam Yokhesh, Leigh M. Levinson, Manuel Castro, Marcos Ribeiro Pereira Barretto, Maurizio Sibilio, Ming Zhang, Noelia Fernandez, Ramon Barber, Randy Gomez, Rune Baggett, Ruoyu Sun, Samiullah Paracha, Sarah Cohen, Selma Sabanovic, Selma Šabanović, Shuo Zhang, Steffi Paepcke, Vera Pereira-Barretto, Xi Yue, Yuan Yuan

First published in London, United Kingdom, 2025 by IntechOpen
IntechOpen is the global imprint of INTECHOPEN LIMITED, registered in England and Wales, registration number: 11086078, 167-169 Great Portland Street, London, W1W 5PF, United Kingdom

For EU product safety concerns: IN TECH d.o.o., Prolaz Marije Krucifikse Kozulić 3, 51000 Rijeka, Croatia, info@intechopen.com or visit our website at intechopen.com.

British Library Cataloguing-in-Publication Data
A catalogue record for this book is available from the British Library

Current State and Future Perspective in Human-Robot Interaction
Edited by Constantin Voloşencu
p. cm.
Print ISBN 978-0-85466-407-8
Online ISBN 978-0-85466-406-1
eBook (PDF) ISBN 978-0-85466-408-5

If disposing of this product, please recycle the paper responsibly.

Meet the editor

Prof. Dr. Constantin Voloşencu graduated as an engineer from Politehnica University Timişoara in Romania, where he also obtained a doctorate. He is currently a Full Professor in the Department of Automation and Applied Informatics at the same university. Prof. Voloşencu is the author of 10 books, 7 book chapters, and more than 170 papers published in journals and conference proceedings. He has also edited 16 books and has 27 patents to his name. He is a manager of research grants, editor-in-chief, and member of international journal editorial boards. He is also a former keynote and plenary speaker, a member of scientific committees, and chair at international conferences. Prof. Volosencu is a science ambassador of LiveDNA and a member of the Asian Council of Science Editors. His research areas include control systems, control of electric drives, fuzzy control systems, neural network applications, fault detection and diagnosis, sensor network applications, monitoring of distributed parameter systems, and power ultrasound applications. He has developed automation equipment for machine tools, spooling and wire drawing machines, high-power ultrasound processes, and more.

Contents

Preface

This book is dedicated to the study of human-robot interaction, a multidisciplinary field that integrates knowledge from robotics, artificial intelligence, natural language processing, psychology, and ethics. The primary goal of this research is to ensure safe interaction with robotic systems, a requirement that traces its origins to the history of robotics and Asimov's Three Laws of Robotics.

Various types of robots have been developed, including social, medical, industrial, educational, mobile, search and rescue, space, and agricultural robots. Social robots assist the elderly, care for children, and support individuals with disabilities. Medical robots play a role in rehabilitation, elder care, and companionship. Robotics has numerous applications, including autonomous robots, gesture recognition, motion planning, personal robots, and robot teams.

Modern technologies such as artificial intelligence, face detection and recognition, human-robot interaction, interactive systems engineering, telematics, and natural language understanding are essential in robot development. Contemporary autonomous systems include localization and mapping technologies that enable intelligent robot movement, natural language processing and generation, and psychological behavior modeling. Anthropomorphic robots mimic the human body structure with biomimetic behavior, while collaborative robots (cobots) are used in industrial manufacturing.

Human-robot interaction must meet highly complex requirements, appearing almost limitless at first glance. Robots should be safe, intuitive, and capable of natural communication through speech, gestures, and facial expressions. They must possess perception and understanding capabilities to recognize and categorize objects, locate humans, and interpret emotions using advanced communication skills. These abilities rely on various methods, including sensor-based human perception, color detection, human kinematics, 3D vision models, speech recognition, neural network architectures, and learning algorithms.

Robots also employ motion planning methods in dynamic environments, ensuring collision-free trajectories for their arms and legs. The practical implementation of artificial psychology is one of the greatest challenges in human-robot interaction, requiring cognitive models and insights from the theory of mind. Research is advancing toward incorporating perception and emotions into robots, necessitating extensive experimental studies. Additionally, the ethical and psychological principles underlying artificial intelligence development remain critical considerations.

This book consists of 10 chapters grouped into five sections, covering key areas of human-robot interaction:

Section 1: Social Robots

Chapter 1 synthesizes findings from extensive studies on older adults' needs at home, identifying requirements for robot-based social assistance and human-robot interaction.

Chapter 2 presents a large-scale study on how older adults perceive and accept assistive technology, contributing to the development of gerontechnology.

Chapter 3 explores critical challenges in long-term child-robot interaction, based on a study conducted in multiple households using a child-oriented family robot simulator. It provides insights into user engagement, verbal interaction analysis, and design recommendations for future child-robot platforms.

Chapter 4 examines general requirements for human-robot interaction in social robots, identifying core components necessary for fostering social awareness in robotic design.

Section 2: Mobile Robots

Chapter 5 discusses deep learning and machine learning-based modeling techniques for scenes and objects in shared human-robot environments, enabling navigation and manipulation tasks.

Section 3: Technology

Chapter 6 explores interactive digital twins and holographic rendering as tools to enhance emotional engagement in learning environments and digital museum experiences.

Chapter 7 presents a study on detecting danger in public spaces through emotion recognition and weapon detection.

Chapter 8 discusses advancements in silent speech interfaces based on neural activity, including the detection of neural and muscle signals through emerging techniques.

Section 4: Machine Ethics

Chapter 9 examines ethical considerations in AI-based robots based on an experimental AI system and user survey data.

Section 5: Medical Robots

Chapter 10 introduces robotic systems designed to improve the assessment, treatment, and rehabilitation of patients with neurodegenerative disorders, highlighting limitations that future research could address.

The editor extends sincere gratitude to the contributing authors for their outstanding work and patience throughout the editorial process. Each chapter was selected through a rigorous review process. Given the authors' expertise, this book is expected to gain recognition among specialists in the field.

Additionally, the editor expresses appreciation to the entire editorial team involved in the book's publication.

Constantin Volosencu
Department of Automation and Applied Informatics,
"Politehnica" University Timisoara,
Timisoara, Romania

Section 1

Social Robots

Chapter 1

Perspective Chapter: Uncovering Older Adult Needs – Applying User-Centered Research Methodologies to Inform Robotics Development and a Call to Action

Katherine M. Tsui, Sarah Cohen, Selma Sabanovic,
Alex Alspach, Rune Baggett, David Crandall and Steffi Paepcke

Abstract

Aging society is a worldwide crisis that began in Japan (JP) and was followed by many more countries, including the United States (US). With this increase in the Older Adult (OA) population, it is pertinent to understand what OAs want for themselves and need to independently live in their own homes for as long as possible. This chapter catalogs our research between 2016 and 2023 about the needs of OAs in and around their homes. Using 10 user-centered research methodologies, we took a cross-cultural approach to conducting 69 studies in the US and JP. The primary goal of these studies was to identify the challenges OAs face in their daily lives and better understand user preferences for robotic assistance for such challenges. This grounded understanding is necessary to design both the robot and the interactions between the human and the robot. Our findings indicate five overarching themes about OA challenges: mobility and stability; moving heavy objects; dexterity; cognitive aging and social support; and sensory and physical decline. This chapter should be used as a guide to inspire the development of robotic technologies that OAs need and want to use and enable them to live independently longer.

Keywords: user-centered research, human-robot interaction, older adults, user need finding, robots, United States, Japan, in-depth interviews, in-home ethnographies and walkthroughs, contextual inquiries, expert interviews, focus groups, participatory design, design sprints, surveys, data sets, diary studies

1. Introduction

The number of Older Adults (OAs) will drastically increase over the next few decades [1, 2]. This is inevitable, universally known and undeniable, and robotics may be one solution to this aging society crisis. Despite best intentions, "use-inspired research" does not often translate into actual benefit for the target population; more commonly, researchers invent tasks that they think will help OAs and that also just happen to be possible with their specific robot or method. The research community needs grounded knowledge of the type of help that OAs want, and then to apply that knowledge to robotics in order to promote OAs to care for themselves. Between 2016 and 2023, Toyota Research Institute (TRI) conducted 69 user studies that involved OAs in the thought and design process to better understand their true needs and wants for aiding independent living. We aim to find opportunities to amplify and uplift aging society, by providing support so that OAs can continue to do the tasks they enjoy and benefit from, and offload tasks that OAs no longer can or care to do. Additionally, we conducted 2 aging-in-place robotics projects - Punyo, a soft robot that looks at OAs' needs relating to physical assistance in the home, and IRIS which focuses on helping OAs to achieve and maintain their sense of purpose, *ikigai*.

Our research focuses on the experiences of both Japanese and American OAs and those who care for them, allowing us to better understand OAs' challenges in day-to-day life. Conducting research in both countries helped us learn about the rich cultural contexts and rituals that researchers and designers should keep in mind when designing human-robot interactions. We conducted comprehensive exploratory research using user-centered design and research techniques that were both technology-agnostic (e.g., generative research that avoids presenting a solution) and also concept-led (e.g., introducing robotic concepts and gauging responses) to verify our understanding.

The world's rapidly-aging human population, in addition to the younger people who will need to care for them, makes it critical to investigate the needs of the aging population. Our perspective is that OA needs are not being met because of a gap between the technology that is being developed and what OAs actually need. Research on aging populations is pertinent as we seek out ways to help our aging population be more independent and successfully age in place [3, 4]. Further, our empirical approach centers on going to see for ourselves, *genchi genbutsu* (現地現物), and understanding what it means to be a care receiver and a professional or familial caregiver by observing and listening in-situ. We see and hear both what is said and shown *and* what is omitted, consciously or otherwise; we have found that some OAs make a series of very slight adjustments over a long period of time and either do not recognize or realize that they are compensating for an issue, or resign themselves to withstand their situation. Our user-centered research takes the perspective of *empowering OAs to care for themselves*, instead of being a person to which care is administered or given. This document provides insights to the user through which the research community can tackle robotic opportunities grounded in real user needs that will positively impact human lives.

2. Background

2.1 The global demographic transition

A large societal impact occurs when a country has more older people than people to take care of them, which is known as a demographic transition [2]. The global demographic transition began around 1800 when there was a *longevity increase* due to the introduction of vaccines (i.e., smallpox) and a decrease in the spreading of deadly contagions [2]. Later, *fertility* began to decline, as parents decided to have fewer children [2, 5, 6]. With the decline in fertility, the working population grows as the population as a whole begins to age. *Population aging*, the final stage of the transition, can occur over 40–50 years [2] and is indicated by the number of people aging out of the workforce growing faster than the number of people joining the workforce [2, 7, 8].

No country in the world has fully completed the demographic transition [2], but Japan was the first to be impacted by it. The JP population is already older than the US one and it is aging faster than the US, which can partly be attributed to the improvements in living standards and medical care [9].

2.2 Robots used with older adults

There have been many studies on Older Adults (OAs) interacting with and using robots. These studies often include OAs with disabilities and living in care facilities [10, 11], rather than focusing on OAs in their domestic environments [12–14]. Much of this research utilizes social robots for social support and to aid with social isolation and depression by providing companionship and emotional support [13, 15, 16]. Whelan et al. assembled a literature review that spanned 44 studies that explain the acceptability of social robots to OA populations [17]. Artificial intelligence and robots can also work together to promote independent aging as they can monitor health, promote social interactions and provide assistance to OAs [4, 18].

There are many activities of daily living (ALDs) that become physically difficult as we age, such as transfer motions from sitting to standing, or reaching high or low to retrieve items [19]. To compensate stability, balance, and mobility, OAs often adopt other methods of assisting with their transfers or tasks [20]. Physical assistant robots can aid OAs by providing support and stability for ADLs [21], completing tasks (i.e., chores) [19, 20], and providing rehabilitation after illness or a medical event [22]. Using a robot as a walker or 'walk-helper' allows users to initiate the movement [21] and can slow physical decline, increase physical activity [23], and decrease fears of injury when walking [24].

As we age, people lose the ability to complete ADLs such as bathing or toileting, and our need for care increases [25]. Issues with mobility and ambulation lead to a major loss of independence in OAs [26]. For many OAs, losing their dignity is one of their largest fears [27], leading them to favor staying at home as they age [28–32]. Various advanced technology can assist this goal of "aging in place," including robotics. Robotics can potentially offer a brighter future for OAs by assisting with physical and mental health problems, providing them with more independence and comfort as they age in place.

2.3 Research methodologies

Between 2016 and 2023, we used 10 user-centered research methodologies to understand what Older Adults (OAs) need and want to independently live in their own homes for as long as possible. We provide a high-level overview of the 10 user-centered methodologies including the benefits of each one as well as our number of participants (*N*) and hours dedicated to each (*Hours*). The various methodologies below are abbreviated and paired with numbering for references to each project we conducted (i.e., *(ID-1)* is In-Depth Interview 1); corresponding study information can be found in Appendix A1 and [33].

An *In-Depth Interview* (ID) is a conversation that can occur in person or virtually (i.e., video call) between a facilitator and a participant. The conversation is a series of questions that the facilitator poses to the participant. The benefit of in-depth interviews is the rich qualitative data that comes from the content flow and style of the questions. Facilitators can dive deeper into a participant's responses and reasoning rather than sticking solely to a script. An interview session generally lasts between 1 and 2.5 hours and typically has a small sample size. We conducted 16 studies with this method; $N = 243$ and $Hours = 277.1$.

There are 2 complimentary methodologies for an in-person interview, In-Home Ethnographies and Walkthroughs and Contextual Inquiries. *In-Home Ethnographies and Walkthroughs* (IHEW) allow researchers to study people in their environments and gather data about their motivations, habits, and needs. This qualitative research method requires the researcher to become immersed in the user's world. Specifically, to observe participants' challenges in their exact environments. The fundamental benefit of IHEW is that researchers gain firsthand knowledge of participant needs. *Contextual Inquiries* (CI) is a method used to further understand a participant's thinking. A user performs their natural actions in their environment while a researcher observes and asks questions "in context" to understand *why* and *how* a user is doing what they do. Sessions can last 2–2.5 hours, have a fairly small sample size, and have a high level of logistical coordination. We conducted 3 IHEW studies with $N = 24$ and $Hours = 57$, and 2 CI studies with $N = 17$ and $Hours = 27.5$. A different type of interview is the *Expert Interview* (EI), in which experts in their given field are interviewed. These interviews are often coupled with in-depth interviews (ID) or contextual inquiries (CI) with the target audience (OAs). The fundamental benefit of EIs is the Subject Matter Expert (SME) supplements the researcher's understanding of a topic. We conducted 9 studies with SMEs; $N = 43$ and $Hours = 94.5$.

Focus Groups (FG) bring together a group of participants to understand various attitudes towards a topic, question, or product. The facilitator guides participants in the discussion to answer and explain their thoughts and experiences. Focus groups can contain between 3 and 10 people, which yields a diversity of opinions in a short period, usually 1–2 hours. The fundamental benefit is that participants can build off each others' responses; this however might limit the amount of individual thought. We conducted 16 studies with this method; $N = 447$ and $Hours = 196$. *Participatory Design* (PD) builds on the Focus Group (FG) methodology to include potential users in the design process, rather than the technology being designed *for* them. PD is sometimes referred to as *co-design* because everyone, including the potential users, has equal voice. PD centers on the lived experiences of end-users to ensure that real-world user needs, feedback, and expectations are reflected in the final product; this is the fundamental benefit. We conducted 2 PD studies; $N = 52$ and $Hours = 39$. The *Design Sprints* (DSP) is a time-bounded structure that allows teams to brainstorm, prototype, and validate a concept or mock-up with user testing [34]. Design Sprints (DSP) have some similarities with Participatory Design (PD) because Subject Matter Experts (SMEs) and potential users might be included on the sprint team. Each sprint team member takes on a role and sticks to a tight schedule to answer a "How Might We" (HMW) question [34]. A HMW question reassembles problems as opportunities and breaks them down into smaller, more manageable pieces. This allows the sprint team to focus on user needs and problems, rather than jumping straight to solutions. We conducted 4 studies with this method; $N = 21$ and $Hours = 83.5$ hours.

A *Survey* is a set of structured questions to which participants self-report information based on their knowledge and lived experiences. Surveys provide quantifiable information easily and quickly. There are 2 types of content: closed-ended survey questions have a limited number of choices for a participant to respond with (e.g., rating scales, choosing options), while open-ended survey questions allow users to include as much or as little information as they would like. Surveys can be used to establish the scaled impact of research findings with a low number of participants. We conducted 9 studies with this method; $N = 5070$ and $Hours = 1752.45$. A *Data Set* (DS) can build on the Survey methodology. Researchers can ask participants *en masse* to collect empirical data on research questions regarding items within their homes or daily life tasks. Participants provided data on measurements, brands, and items in cabinets. Data Sets can be followed up on with an in-depth interview for qualitative insights. We conducted 3 studies with this method; $N = 76$ and $Hours = 58$. Similarly, *Diary Studies* (DIARY) are a longitudinal research method that asks participants to self-report behaviors, thoughts, or activities over time. A survey can be used to collect this data. This can range from just a few questions each day and extend to a few days, a month, or more. We conducted 2 diary studies; $N = 30$ and $Hours = 20.15$.

3. Older adults user needs assessments: Generational research in JP and the US

Toyota Research Institute's vision is to dramatically improve the quality of life for individuals and society [35] by better understanding Older Adults (OAs) as whole people to better support them in their independent living. To do this, we took a comprehensive, empirical approach. We conducted 69 studies with OAs totaling in 5990 participants and 2605.2 hours of data collected using 10 different research methods (See details in Tables A1-A3, and summaries and key findings for individual user studies in [33]). Using a variety of research methods allowed us to better understand the breadth and depth of OA challenges and their desires for support and to ensure that this research contributes to developing worthwhile robotic capabilities. The primary goal was to understand opportunity areas, ranging from mobility to cooking and beyond.

This research was conducted across populations of rural and urban JP and US[1] OAs; we define OAs as people aged 65+. In addition, investigations with younger generations in JP highlighted generational[2] differences, revealing unique user needs and home challenges specific to JP. We found distinct needs and desires depending on the generations in which participants fell.

One commonality across all populations is that OAs want to remain independent and do things for themselves for as long as possible, even lifting heavy objects as they "know their limit" *(ID-1)*. It is crucial to point out that many caregivers are family members (i.e., an adult child or spouse [36]). In JP a common caregiver and care receiver dynamic is identified as as "elderly caring for the elderly" [老老介護 *rou-rou kaigo*] [37].

Still, there are striking cultural differences between OAs in JP and the US. OAs_{JP} do not want to be a burden to their family or others, physically or financially. Thus, OAs_{JP} view physical and mental fitness as their new job after retirement and often re-enter the workforce to aid in their fitness [38, 39]. Despite strong trust in long-term

[1] US Generations: Baby Boom Generation (Born between 1946 and 1964); Generation X (Born between 1965 and 1980); Millennial Generation or Generation Y (Born between 1981 and 1996)

[2] JP Generations: Dankai (Zenkyōtō) 1947–1949; Shirakē (Dansō)1950–1964; Bubble 1965–1971; Ice age (Lost generation) 1971–1982; Pressure 1983–1987; Satori 1987–2004.

care services, it can be difficult for OAs_{JP} to admit that they are aging and get the services that they are guaranteed *(FG-2)*.

Section 3.1 is a review of the in-home challenges and needs of OAs, that we uncovered in our empirical research; summary and key findings for individual user studies are provided at https://osf.io/8729k [33]. Findings were pulled from researcher notes and transcripts, then analyzed by affinity diagramming.[3] Researchers organized the many individual notes into groups, creating *affinities*. They added titles to these groupings, which developed into themes of aging. Researchers tracked and sought out recurring themes across individual reports surrounding challenges and support received or desired by OAs. This book chapter synthesizes the individual studies mentioned, pulling from the research reports, raw data, and key insights. We illustrate important themes through chosen quotes by thematic analysis [40].

3.1 The 5 main challenges of aging

As people age, they experience more challenges physically and cognitively. There are 5 main aging-related challenges that Older Adults (OAs) face and that may benefit from robotic support: (1) Mobility and Stability; (2) Large and Heavy Item Manipulation; (3) Dexterity; (4) Cognitive Aging and Social Support; and (5) Physical Aging and Sensory Decline. These themes are consistent with prior research on challenges that OAs face as they age [41–44]. Understanding OAs' individual limits and desires for support can allow us to create solutions that work in and outside of the home to help them prevent injury and to continue to live independently. Caregivers may respond to challenges by providing physical support to their care receiver while they perform the task themselves or take over tasks from their care receiver. Ultimately, if an OA's level of needed care is beyond what a familial caregiver can provide, familial caregivers may move their OA into assisted living.

3.1.1 Challenge #1: Mobility and stability

Challenges of walking *(IHEW-1, SURVEY-4, IHEW-2)* climbing stairs *(IHEW-3)*, navigating cluttered spaces *(EI-2)*, getting in and out of a bath or low position *(IHEW-2, IHEW-3)*, and bending low and reaching high *(IHEW-1, IHEW-2, IHEW-3, SURVEY-4)* begin to slow OAs down. While some OAs see slowing down as *"just a part of getting older,"* personal mobility decline can cause OAs to adopt potentially unsafe behaviors such as leaning or pushing off of unstable furniture *(IHEW-2)* when assistive devices are not readily available *(IHEW-1, SURVEY-1, SURVEY-4, IHEW-2)*.

Stairs and clutter. In 2020, 14 million (27.6%) OAs_{JP} reported falling [45]. In 2021, 38,742 (78.0 per 100,000 population) OAs died as a result of unintentional falls [45]. Stairs can make an area completely inaccessible. Many JP homes have steep, tall staircases to upper levels and even one or two steps up to other rooms; 47% of US homes have stairs inside the home [46]. One participant cannot go to any restaurant with steps and no elevator access, and another OA does not clean upstairs anymore because the stairs are too strenuous *(IHEW-3)*. Clutter is an inescapable aspect of human life that compounds challenges to mobility [47] as it leaves less space to navigate and

[3] https://www.nngroup.com/articles/affinity-diagram.

creates a tripping hazard, therefore posing a safety risk. Because urban JP homes have little storage, items may be stored on the steps of their steep staircase *(SURVEY-6)*.

Home modifications. Some OAs have handrails and grab bars added to areas near the bed, toilet, and stairs. One study highlights how the accessibility needs changed within a home for one OAs_{JP} *(IHEW-3)*. He had previously installed grab bars in his home for his aging parents but removed them when they passed away because the bars were an eyesore. He ultimately added the grab bars back to his home as they became more necessary in his older age, demonstrating how mobility and stability change over time.

Bathing. In the event of needing help, bathing and personal care are particularly sensitive topics. Most bathroom designs are not inherently accessible. Some require a high step up into a tub or shower *(IHEW-2)*, while water creates slippery surfaces which can lead to increases in falls [47–49]. An OA may need to use grab bars and/or require their caregivers' help to get up from a seated bathing position *(IHEW-2, IHEW-3)*.

Stepping and transfers movements. Struggles with climbing stairs [47, 48, 50] and sitting to standing [51] can indicate that more intense care is needed, and OAs may need to eventually move to care facilities *(SURVEY-1, IHEW-3)* [52]. One OA couple struggles with standing up from a deep couch. Only one will sit on the sofa at a time, and only if the other is present to help them back up *(IHEW-1)*. One OA uses a stool to sit on while weeding in the garden so that she does not have to bend and crouch as much *(IHEW-3)*. Reaching to get items out of high cabinets or placing things onto altars requires a chair or stool *(IHEW-3)*. One OA stated that "*it's getting scary to use a ladder to climb up to my roof; it used to be nothing when I was younger*" *(IHEW-3)*.

3.1.2 Challenge #2: Moving heavy objects

Moving heavy items *(SURVEY-2, ID-1, CI-1, IHEW-2, IHEW-3, DIARY-1)* may not seem like a regular occurrence, but individuals have differing definitions of 'heavy items' based on their individual physical strength. OAs defined "heavy" by giving examples of objects that present a challenge to them (e.g., large bottles of olive oil, a bag of sports equipment, groceries, or a bag of rice) *(ID-1, CI-1, IHEW-2, IHEW-3)*.

Groceries. Outside of the home, a frequent task that often requires heavy lifting is grocery shopping. American OAs use a grocery cart and get support from baggers to take groceries to the car *(CI-1)*. One OA with severe osteoporosis reported being unable to move anything heavier than 2 pounds and must ask a delivery driver to come inside and transfer some olive oil from a large container to a smaller container that she can lift *(IHEW-1)*. In JP, grocery shopping happens nearly every day to obtain the freshest ingredients and is seen as a "*reason to go out*," indicating that OAs can still lead independent lives. Grocery carts are rarely used if someone is going for just a few items; instead, a basket is used or items are carried in hand. OAs_{JP} shoppers may limit their purchases, buying many small items or one large item but not both during the same trip *(IHEW-2)*, or may opt grocery delivery *(IHEW-2, ID-8)*.

Chores and hobbies. Around the home, OAs_{JP} struggle with cleaning tasks such as vacuuming, and some leave it for another family member to complete *(IHEW-2, ID-8, ID-7)*. OAs_{JP} prefer to dry their laundry outside, which can mean carrying a heavy basket of wet laundry up steep stairs to the balcony *(IHEW-3)*. An inability to move heavy items also impedes social activities. One OA cannot carry her heavy painting supplies, so she no longer attends her class *(IHEW-2)*.

3.1.3 Challenge #3: Dexterity

Applications that require dexterity are embedded into everyday tasks. Declining fine motor skills and hand strength may not only slow an OA down and prevent them from doing what they need and want to do, but can cause serious injury if the OA attempts a task that they are unable to complete. Losing grip strength or the ability to firmly grasp can put an OA in a dangerous situation. Grab bars and accessibility features added to the home to support mobility and stability are not useful if the OA is unable to grasp them, as in the case of one OA with partial paralysis *(IHEW-2)*. Similarly, difficulty with grasping can cause safety risks when carrying items, such as removing hot food from the microwave or carrying hot coffee *(IHEW-1)*.

Opening containers. Opening packages and medication containers can be so challenging that a manager at an independent living center said this one of the most common requests at the front desk *(IHEW-1)*. OAs living alone do not always have help available, particularly in more rural areas. One isolated OAs_{JP} opens bottles and jars using a silicone gripper tool, citing a lack of strength and a hand fracture *(IHEW-3)*.

Dressing. Dressing oneself, an Activity of Daily Living (ADL), can become difficult with diminished dexterity. One OA who has trouble dressing and undressing suggested that she wants to get pants with fasteners up the side *(ID-7)*. This challenge is compounded when an OA is unable to balance. One OA found it hard to reach to put her socks on and cut her toenails; she has stopped changing clothes often *(IHEW-2)*.

3.1.4 Challenge #4: Cognitive aging and social support

Cognitive impairments come with a stigma and impede Older Adults' (OAs) ability to both perform tasks and to maintain their social lives. There is a pervasive fear of *dementia* in JP, and continuing to do (even heavy, physical) tasks as long as possible is seen as a crucial tool for maintaining mental abilities *(IHEW-2, ID-15, IHEW-3)*. OAs_{JP} seek these out by, for example, participating in Adult Day Care activities like math and word searches *(IHEW-3)*, doing chores that have multiple steps, or shopping for groceries. Some OAs enjoy grocery shopping because it is a goal-directed reason to leave the house and socialize *(ID-14)*. Volunteering or a post-retirement job can provide mental stimulation, an opportunity to socialize, and a sense of purpose (*ikigai*); many OAs_{JP} noted that they find meaning in their lives by serving others *(IHEW-2)*.

OAs want to maintain their physical and mental health to stave off senility because they have watched their parents experience dementia and fear this cognitive decline *(ID-14, ID-15)*. With memory loss and other cognitive declines, *socializing* becomes challenging *(IHEW-1)*. One OA enjoyed Mahjong but often forgot the rules *(IHEW-2)*, while another OA is no longer able to recognize her friends *(IHEW-2)*. One OAs_{JP} noted, "*I have friends nearby, but if I stop driving my car, I am afraid I'd suffer dementia... It's like heaven to be able to drive*" *(IHEW-3)*. The act of driving a vehicle requires a person to have both cognitive fitness and physical coordination and provides the means to leave the house and socialize.

3.1.5 Challenge #5: Sensory and physical health

Physical aging and sensory declines in vision and hearing can also negatively affect OAs *(IHEW-1, IHEW-2, IHEW-3)*.

Sensory decline. Decreasing *hearing* ability can limit OAs' social interactions *(IHEW-1, IHEW-2, IHEW-3)*. One caregiver explained that they have stopped trying to have conversations with their care receiver due to the care receiver's hearing difficulties *(IHEW-2)*. It can be especially hard to converse in noisy environments; one OA stopped attending town meetings because she could not hear well *(IHEW-3)*, and others had to get words repeated or shouted at them *(IHEW-2)*. This leads to overall difficulty in attempting to attend social events and can lead OAs to completely miss out on these interactions. Beyond hearing problems, declining *vision* significantly impacts day-to-day activities. Vision affects the ability to drive, especially at night. Small font sizes on price labels at the grocery store can prevent OAs from being able to purchase the items they need or want because they cannot read the label *(IHEW-2)*.

Fitness. While it is challenging to stave off sensory decline, fitness is something that OAs can use to maintain their abilities. One OA treats going to the gym like a job (going 4 times a week) and a place to socialize *(IHEW-2)*. Another OA began a daily bath time exercise in which she sat on her knees and raised and lowered her body 100 times, as the activity did not hurt her in the water; this strengthened her knees and made them more flexible *(IHEW-3)*. Maintaining physical fitness can benefit coordination, strength, and social connections. OAs can also use physical fitness to set goals and create a sense of purpose.

3.2 Specific application areas

We have discussed 5 main challenges that Older Adults (OAs) face, yet solutions that support these challenges in perfect lab conditions may not address the contextual situation in which the challenges occur. For example, a robot may be able to retrieve a heavy box off a high shelf, but in dim lighting, in a small kitchen, with clutter in the way, the task increases in complexity. When considering opportunities for robotic applications, researchers and designers need to consider the situations in which these challenges appear (i.e., creating a robot that can bend down low or support an OA to bend is not enough). For example, roboticists may create a solution for the problem "*reaching is hard*." However, the solution may not be robust enough to solve the contextualized problem of "*reaching into a deep washing machine and pulling out wet, heavy clothing is hard*." Addressing the tasks and activities in which bending is involved allows people to live and thrive independently. By closely evaluating specific application areas, one can see the interactions between the environment and the many challenges that might arise when completing the task. The following sections will walk readers through 3 specific scenarios that participants found challenging: (1) Cleaning, (2) Cooking, and (3) Laundry.

3.2.1 Application #1: Cleaning

Cleaning the house is a repetitive, near-daily task that many Older Adults (OAs) struggle with or would like to offload *(IHEW-1, IHEW-2, ID-8, ID-6, IHEW-3, ID-7)*. Cleaning the floor can be challenging due to the OA's mobility, stability, and/or ability to bend low. An OA must first bend down to pick up clutter before using tools to clean the floor. The process involves not just standing and bending but also setting up tasks like walking around with a heavy vacuum *(ID-7, IHEW-1, ID-6, ID-8, IHEW-2)*.

Washing the floors is especially challenging on one's hands and knees, and OAs often choose not to do this. One OAs_{JP} who lives with her son and his family does not wipe the floors by hand anymore: "*I used to clean the floor by myself, but it's hard to bend and squeeze out the rag, so my daughter-in-law got [a robot vacuum] for me*" *(IHEW-3)*. Another OA who struggles with bending can no longer get down to scrub. Instead, she uses a grabber tool with a rag to clean the floor; she changes the rag out to clean the toilet *(IHEW-2)*.

Some OA caregivers must balance cleaning with other responsibilities such as working and taking care of young children, which often means doing the cleaning, cooking, and home tasks for two households. Gender roles can also play a part in cleaning; husbands have been unaware of the immense amount of work necessary to keep the home clean, which wives primarily do *(ID-7, ID-8)*.

Respondents across multiple studies—generational age groups and countries—state that cleaning is repetitive and time-consuming. In a cleaning survey, more than half of JP respondents tidy more than once daily, focusing on books, newspapers, and children's toys *(SURVEY-6)*. Living rooms in JP are cleaned most frequently—up to 3 times a day—because of the high traffic. In the US, similarly, cleaning the kitchen countertop, stove, and refrigerator is the highest priority *(ID-6)*. Cleaning the bathroom is essential for keeping a space hygienic since all household members bathe daily. In bathrooms, high humidity causes mold to form quickly, which is unsightly and unhygienic; soap scum and hard water stains require intensive scrubbing. When asked what they disliked about cleaning, one participant answered, "*Cleaning the bathtub. You have to use your full force to scrape the surface and wipe it all.*" Except for a few individuals, most OAs dislike cleaning, have a hard time with the repetition of it, and feel that some tasks should be easier than they are *(SURVEY-6)*.

3.2.2 Application #2: Cooking

Challenges with cooking arise across the whole process: acquiring ingredients, preparing food, cooking, and even eating. In JP, kitchens are much smaller compared to American kitchens *(EI-3)*, and clutter can further shrink an already minimal workspace *(ID-7)*. Due to a focus on freshness and variety, many Japanese Older Adults (OAs) grocery shop nearly daily.

The physical strain of carrying purchases home limits what OAs choose to get. Grocery shopping was identified in multiple studies as the task most often supported by a caregiver (most notably, 100% of caregiver participants interviewed in FG-3, and 54% of self-identified caregivers from a general population online survey *(FG-3)*). According to a 2019 JP survey, half of all shoppers buy groceries 3–4 times a week, and 30% of women aged 60 and older shop more than 5 times a week *(EI-3)*. Attention is placed on fresh food and only preparing the quantity that will be consumed for the meal. Leftovers are not common in JP households.

Cooking fresh meals is time-consuming *(IHEW-2, ID-8)*. Young parents$_{JP}$ struggle with keeping meal ideas fresh to prevent boredom *(ID-8)*. Younger and older generations alike utilize pre-prepared foods, like those found at a supermarket or a convenience store, despite that these meals may not be as nutritious as desired. One OA only eats convenience store food. Her diet is mostly *soba* and *udon* noodles, with no vegetables, meat, or fish *(IHEW-2)*. The wife of another OA created a chart for her husband that indicated how much of each type of food he should be eating. When asked if he uses it to cook, however, he said, "*I don't look at it that much*" *(IHEW-3)*.

People who derive purpose from creating meals want to continue to do so as they age. Moving heavy pots or standing for long periods of time an be challenging for OAs; some have a stool near the stove. Chopping takes time and precision, and hard foods like pumpkins can become challenging for OAs. One OA used her whole body weight to cut an apple, which could be dangerous if she were to slip with a sharp knife *(IHEW-2)*. Peeling vegetables is also laborious and requires dexterity *(EI-3)*.

Robotic support can assist OAs in maintaining traditional food preparation while also offloading dull or repetitive tasks; this allows OAs to remain in charge of the task while also enabling them to cook safely. In one study, participants were shown an animated video where the tiring, repetitive work was offloaded for an OA with arthritis. Robotic arms watch an OA shave bonito flakes, and then the robot completes the task by itself. This resonated well because participants liked that they could choose to do it themselves or ask for help from the robot *(FG-3)*. OAs expected the robots to exert the right amount of pressure and precision to prepare ingredients to their liking.

Cooking is one way to maintain mental acuity *(ID-8)*; one OA feels that she must cook regularly to not forget this skill. However, cooking can be a safety hazard for those with memory impairments *(IHEW-2, EI-3)*. Although the son (caregiver) circled the microwave buttons his mother should use, she almost caused a fire by using the wrong settings because of dementia *(IHEW-2)*.

Kitchens in JP often are only big enough for one person to use at a time *(DS-1)*. Combined with a lack of storage space and clutter, having a clear cooking surface is a struggle. One OAs_{JP} said, "*There are so many dishes that I need to wash, and if I go a while without washing, the sink would become full, and the dishes would pile up. That is the biggest worry for me: when I cannot continue with my cooking or I cannot cook the rice because the sink is full, and I cannot wash the rice cooker ... It's washing the dishes that is the biggest challenge for me. I already have a dishwasher, but I still have to put all the dishes in the machine manually by myself. When you go to a sushi place, they often have the sushi on a conveyor belt. I wish I could put the dishes on such a thing and have the dishes automatically wash and dry*" *(ID-7)*. Chores like tidying and cleaning make cooking even more daunting.

3.2.3 Application #3: Laundry

Laundry is an application that involves many of the underlying challenges related to aging. In a study in which an animation shows a robot performing laundry tasks, 59% of the JP general population participants expressed that they would like to offload doing laundry *(SURVEY-5)*. We observed that Japanese Older Adults (OAs) were more meticulous about cleaning laundry than those in the US; regardless, laundry is still a never-ending, arduous task, and the physical challenges are taxing for OAs.

Washing machines in JP tend to be small, so it is common for 1–2 person households to run a load 2–3 times per week, and for 3+ person households to run 4–7 times per week. It is time-consuming and troublesome to sort dirty laundry into loads for washing. Prewashing stains takes time and effort before placing items into the washing machine. Loading is tricky, as some participants reported that they turn clothes inside out before washing. Detergent containers can be large, heavy, and hard to hold while measuring the soap. When the cycle is done, users must squat or bend to pull laundry out of a front-load machine or reach down into a top-load machine. Wet laundry can be tangled from the washing cycle, and it takes strength to untangle wet clothes. Participants reported that if they do not attend to wet laundry quickly enough, it can get smelly and they must rewash the load.

Many OAs_{JP} do not have or prefer not to use dryers, with only 3% of respondents using them *(EI-6)*; dryers can take a long time, contribute to high energy bills, or damage clothes if too high a temperature is used. Instead, OAs_{JP} indicated that the smell of the sun was an important—or even favorite—redeeming quality of clean laundry, so hanging it outdoors is preferred. In JP households with more than one story, it is common for the washing machine to be on the first floor while the best place to hang dry is the second-floor balcony. OAs_{JP} must carry a load of wet laundry upstairs to the balcony, compounding their challenges of lifting heavy objects and climbing steep stairs; seasonally, they also air out futons, which are traditional bedding and a heavy, bulky object *(IHEW-2)*. Items are individually hung to dry with clips, and care is taken to pat or shake out the wrinkles so that the article dries nicely. Wrinkle removal is the most disliked part of the task, but also the step many put the most care into. OAs_{JP} must raise their arms above their heads to reach the clips. One OAs_{JP} cannot reach up high, so her ability to do laundry is limited: "*I can't hang clothes, so when the helper comes, it's the only time I do laundry. So they can hang the clothes up for me .... [I don't hang laundry outside because] I wouldn't be able to bring it inside, so I usually hang it inside*" as she points to indoor molding *(IHEW-2)*. Folding items takes a significant amount of time, and folding neatly takes additional mental effort to avoid wrinkles. Ironing is another hassle due to the need for setting up and cleaning up *(ID-1, IHEW-1, ID-7, DS-2, EI-6, SURVEY-5, DIARY-2)*.

4. Developing robots to support older adults

Our extensive knowledge about Older Adults' (OAs) needs and wants allowed us to identify promising opportunities to improve their quality of life while aging in place. We describe two robots that serve very different purposes. First, *Punyo* provides physical assistance and *IRIS* focuses on social support; see **Figure 1**. Punyo is a human-sized humanoid robot with a soft, sensing exterior that can pick up heavy,

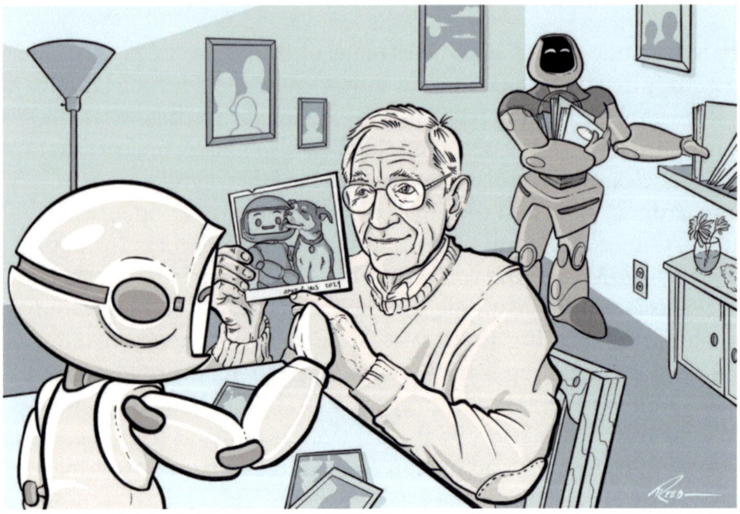

Figure 1.
Concept art of an Older Adult (OA) looking at photos of moments of his life; Punyo gathers and brings over the heavy photo albums, and the OA chooses photos to show IRIS, who asks caring questions about the scene.

bulky, unwieldy objects using its whole body, instead of just its hands. Second, the Interactive Robot for *Ikigai* Support (IRIS) is a tabletop humanoid robot designed to be a social enabler, by helping OAs reflect on what is meaningful in their lives and ways that they might connect with their friends and loved ones.

4.1 Punyo: TRI's soft humanoid research platform

The Punyo Project at Toyota Research Institute (TRI) was established to explore the potential roles and benefits of a soft, capable, friendly-looking humanoid robot as a domestic assistant. Inspired by OA needs, concept art was generated at the outset of this project to inspire and motivate Toyota and TRI leadership and the Punyo team. It was also used to inspire OAs to think deeply about their ability to accept and use such a robot to age safely and happily in place, and illuminate the changes they might like to see to make it even better. We created concept art of Punyo lifting and moving large, heavy objects, tidying up, socializing with family, physically assisting an OA from sitting to standing, guiding exercise, as well as other chores and interactions.

Soft, sensing hardware. The current version of Punyo is a humanoid upper body—a chest, arms, head, and waist but no legs. Punyo's hands, arms, and chest are covered with soft tactile sensors so it can feel contact. The softness allows Punyo to conform to the items it grasps for stable manipulation (**Figure 2**). Its tactile sensors allow Punyo to sense contact and apply controlled forces on objects and its environment, and, importantly, to interact gently with people. The sensors on Punyo's arms and chest have a customizable fabric covering allowing us to change physical properties, color, and style.

Addressing challenge #2: Large, heavy, unwieldy object manipulation. Punyo's main use case is the manipulation of large, heavy objects that require more than just the robot's hands, i.e. requiring its arms and chest. In a domestic environment, where robots co-habitate with people in small, unstructured spaces, it is critical to be safe even in unexpected situations. While the softness of Punyo helps, it is important to keep the mass of the sensors and the underlying rigid components and actuators as lightweight and low-powered as possible [53]. To manipulate heavy objects using less power, a robot must efficiently utilize its motors and body structure—for example, by leaning back and holding objects between its arms and chest, as opposed to using outstretched arms. Synthesizing such intelligent motions for a robot is not computationally straightforward, so the Punyo team has invested heavily into learning methods to achieve human-like, capable, efficient strategies. Punyo is able to grasp, lift, and rotate large objects like boxes and 5-gallon water jugs [54, 55]. More efficient use of the robot's body and actuation power increases payload and battery life and ultimately enables lower-cost, safer systems.

Figure 2.
(Left) Concept art of Punyo pushing open the front door with its arms full. Punyo prototype grasping several yoga mats (Middle) and holding a 5 gallon water jug (Right).

4.2 Interactive robot for *ikigai* support: I.R.I.S.

Researchers from TRI and Indiana University Bloomington worked together with Older Adults (OAs) who were either diagnosed with dementia or caregivers of those diagnosed to design a home robot to support their *ikigai*, or "meaning in life," as they age. Changes in and sometimes loss of a sense of purpose, coupled with a lack of meaningful engagement with the community, can be major challenges of aging. Improving the quality of later life requires maintaining and sometimes increasing OAs' sense of purpose by addressing their need to conduct meaningful everyday activities, maintain interpersonal relationships, and participate in and contribute to society. Through interviews, surveys, observations in a dementia care center, and ongoing Participatory Design sessions with an OA panel (some of whom have dementia themselves, and some are caregivers) [56–58], we developed social robot interaction capabilities and intervention activities that facilitate OAs' reflections on what brings them meaning and provide ways to maintain and strengthen their *ikigai*.

Ikigai (生きがい), loosely translated as one's sense of meaning in life, is commonly understood as essential for individuals to lead a fulfilling life. Previously, ikigai has been described as having three "levels:" 1st person (personal), 2nd person (interpersonal), and 3rd person (community). Individuals may have sources of ikigai at any one of these levels, though it is ideally experienced at all three levels (e.g., personal hobbies, family, and volunteering) [59]. Ikigai is associated with various health benefits, such as prolonged health, longevity, and positive effects on well-being. Recognizing these benefits, the Japanese government incorporates the promotion of ikigai in its policy-making, with OAs as the main target. As a result, Senior Citizens Clubs, local "ikigai centers," and ikigai employment programs were established. While in Japan ikigai as a wellness concept became popularized in the 1960s, it has spread to international audiences more recently; Dr. Akihiro Hasegawa, a leading researcher in *ikigai*, calls the period from the 2000s onwards the "Renaissance of ikigai research" [60].

Addressing challenge #4: Robot as social enabler. Social robots generally are developed to create and build relationships with their users. However, IRIS is different because it is a social enabler, meaning that the goal is to help OAs engage with other people through IRIS. Inspired by studies that show ikigai is an important aspect of wellness throughout the lifespan, and by TRI's pursuit of ways in which robotic technologies can support aging, we worked with OAs in the US and JP to understand how they define their own ikigai and how a social robot might assist them reflect and act on their ikigai. Our goal with IRIS is for OAs to be inspired by their reflection and to suggest ways in which they can connect with members of their family, their friends, and larger community; see Section 3.1.4.

Meet IRIS: Designing robot interactions. We built upon the commercially-available QT robotic platform from LuxAI [61] by developing relevant perception capabilities and personalized ikigai models of and recommendations for participants. We aimed to design human-robot interactions that help OAs achieve increased *feelings of ikigai* by helping Older Adults (OAs) identify and *reflect on activities and relationships* that they find meaningful and suggesting further opportunities for developing meaningful pursuits and social connections. Through interviews and co-design workshops with OAs in Japan and the US, we developed 3 main activities that IRIS performs with older adults: a photo-based conversation; guided reflection on short-term and long-term accomplishments and goals; and an open chat with ikigai-related questions [62]. To support these conversations, the robot uses pre-determined questions interspersed with contextual responses generated by a Large Language Model (GPT3), allowing for dynamically generated follow-up questions and acknowledgments tailored to the user's responses, enhancing dialog cohesiveness and engagement. Based on user interactions, the system creates a user profile that stores activities and relationships the person finds meaningful; these can later be used to guide further conversation or make recommendations.

We evaluated IRIS's functions with 79 OAs ($OAs_{US} = 40$, $OAs_{JP} = 39$) in single-session interactions. We also placed IRIS in homes and eldercare facilities for 2-week field trials; 29 OAs ($OAs_{US} = 12$, $OAs_{JP} = 17$) interacted with the robot repeatedly over multiple sessions. Overall, participants are willing to talk to the robot about topics that matter to them, including things they said they may not talk to other people about. They reported overall feeling heard and engaged by the robot. However, OAs_{JP} shared less detail about themselves; they also provided more detailed feedback on the robot, recognizing its potential role in their lives, while OAs_{JP} had more difficulty imagining using the robot long term. We found that care partners of OAs living with dementia were particularly interested in the possibility of using the robot to supplement their care [63].

5. Implications for home-helper robots

Through robotics, Older Adults (OAs) themselves could contribute to their own care provider network, which is a shift from being the object "being cared for." There is a clear need for creating innovative robotic systems that address OAs' actual needs and enable them to successfully support them as they age in place. These home-helper robots must address the primary challenges that OAs face daily, which include mobility, stability, and social connection, and assist OAs in their routines and their simple yet endless daily tasks and chores (e.g., cleaning, cooking multiple meals, doing laundry; see Section 3.2). The robots must make OAs feel comfortable with receiving help from, using, and inviting robots into their homes. Because OAs only want assistance when they absolutely need it (and not before) and because the type of assistance varies based on task and a user's preference, we propose 3 tenets which we believe that home-helper robots for aging in place must abide by.

First, home-helper robots should not replace human caregivers and should empower care receivers to do for themselves. Ideally, roboticists should not create a less effective or less capable product than a human, and solutions should not carry the stigma of an elderly assistive device. OAs suggested that those with dementia may be supported in areas such as retrieving misplaced items. The robot could know where each item in the home is and instead of directly fetching the item, could support people with cognitive decline by serving as a memory tool to help retrace a user's steps. As noted in Section 3.1.2, defining a "heavy" object is subjective based on each individual OA's physical strengths, and a robot should be aware of the payload maximum that the OA can lift on their own. Instead of always lifting heavy objects for the OA, a robot could instead help maintain an OA's autonomy by lightening larger loads (e.g., decanting a heavy bottle by pouring it into a smaller one) so that the OA can independently lift the smaller amount without the robot's assistance.

Transfer is one of the most frequently occurring Activities of Daily Living (ADL), including lying down to sitting up, sit-to-stand, and ambulation, as noted in Section 3.1.1. The ideal home-helper robot would assist with transfer without relying on another person, thus maintaining OAs' dignity. Such a robot would need to coordinate with and direct the OA to do as much of the initial transfer on their own, and then provide the appropriate assistance to help them complete the maneuver. Unlike a "heavy object," an OA has autonomy and some level of control over their body. The robot must understand *when*, *how*, and *how much* assistance to provide without causing discomfort or injury to the OA. The robot providing only the necessary amount of assistance allows the OA to maintain their skill and confidence.

Second, home-helper robots should personalize and adapt to the OA. It is paramount to understand the OAs' own goals and their expectations for a robot, remembering that every OA is an individual with preferences which can change over time. In the context of Mobility and Stability (3.1.1), as people age, their strength declines, which may necessitate a robot with increasing levels of support over time; the amount of assistance needed can even change throughout the day.

OAs should be able to customize the way in which a robot completes a task and dictate if the task is fully or only partially offloaded. Robots may empower OAs to maintain control by handling only the most challenging or distasteful aspect of the task as assessed by the OAs themselves; for example, some OAs enjoy cleaning and would not want this task offloaded. Robots should support as many accessibility features as possible for OAs to customize. Communication from a robot should take into account declines in senses, and OAs should have the control to increase and decrease volume, screen brightness, etc. (See 3.1.5.) If there is a display, it should be clear and large. A robot could translate hard-to-read text or hard-to-hear audio and serve as a conversational conduit.

Third, home-helper robots should guard OAs from harm. OAs may hide their physical and/or cognitive decline from their loved ones. Robotic solutions for aging in place can be another layer of care management alongside human care and interaction.

In the context of physical safety, it would be beneficial if robots could maintain a safe environment. For example, in clearing clutter from the floor, including picking up items, identifying what they are, and putting them away, the robot maintains clear and safe walking areas for OAs (3.1.1). Designers should define legible interactions by creating robots that move at appropriate speeds and maintain distances in the home when navigating to not become an obstacle itself; large mobile bases are not ideal for a home environment. Ideally, robots will support personal mobility and stair climbing with just-in-time support wherever an OA needs to support themselves.

Concerning cyber safety, many OAs indicated that privacy is an important detail to understand before they adopt robotic technology. The care receiver and caregiver should set boundaries and privacy constraints. There should be transparency about what data is collected, stored, and shared; protections should be put in place to reduce leaks or hacks.

6. Conclusions and a call to action

At Toyota Research Institute, we have demonstrated concrete steps towards solving real problems with real needs to help real people, specifically towards the worldwide aging crisis. Between 2016 and 2023, we conducted extensive user studies, collected 2605.2 hours of data, and gained comprehensive knowledge about the needs and desires that Japanese and American Older Adults (OAs) have as they age from a multi-generational perspective. We discuss 2 projects that we undertook based on this major body of empirical user needs finding research: Punyo can provide OAs with physical assistance and focuses on lifting and carrying heavy, bulky, and/or unwieldy items using its whole body (arms, end effectors, chest), whereas IRIS can provide social support by enabling OAs to reflect on what is meaningful in their lives (ikigai); see 4.1 and 4.2, respectively.

The primary commonality across the US and Japanese populations was that OAs want to remain independent and do things for themselves for as long as possible. OAs can feel ashamed, embarrassed, or like a burden when repeatedly asking for help from loved ones or receiving help from caregivers. Robotic support is a possible solution to this never ending crisis as OAs can freely ask for assistance and do not have to worry about the robot tiring. With the help from assistive robots, OAs could receive the support they need as they age in place, with a robot attending to the OA whenever they want, at any hour of the day, and as many times as they need without feeling like they are bothersome.

We found that there are 5 primary challenges that OAs face in their daily living that can benefit from robotic support: (1) Mobility and Stability; (2) Large and Heavy Item Manipulation; (3) Dexterity; (4) Cognitive Aging and Social Support; and (5) Physical Aging and Sensory Decline. Keeping these challenges in mind, there are several implications for home-helper robots that must be followed. OAs do not want robots to *replace* human caregivers but to *empower* OAs to be as independent as possible. OAs desire robots to be *personalizable* to individual OA needs based on whom it provides cares. Finally, OAs demand that robots not only *do not cause harm to them*, but they should also *protect OAs from harm* and *prevent dangerous situations* from occurring. We share our knowledge to inspire the robotics research community to focus on meeting the real needs of an aging society and to help OAs live independently in their own homes longer, care for themselves, engage with their community, maintain a sense of fulfillment, and live their best lives.

We have only scratched the surface of what is possible, and there is so much more that we as a community should do to support OAs' aging in place. We hope that designers, researchers, and engineers take to heart the importance of *genchi genbutsu*: going to see for themselves and seeking the opinions and feedback from OAs (care receivers) as well as their formal and informal caregivers (healthcare professionals, OA spouse, and/or adult children). There is a natural progression of human aging, and we believe that robots built in response to this call to action can also benefit other users in different parts of the world.

This chapter presented a perspective on research, offered a viewpoint on existing problems with fundamental characteristics of older adults, and discussed the need for implementation of innovative robotic systems. Future advances are necessary on the topic, and we hope this chapter inspires others to work carry on the ideas in this chapter.

Acknowledgements

The authors of the paper wish to thank Leticia Priebe Rocha, Bisi Chikwendu, Dikhia Kaur, Talia Kolikant, Margaret Krupp, Abhi Angrihotri, Chris Gidwell, Toffee Albina, Max Bajracharya, CarterJMRN, Will Knight, Ann Booth-Clibborn, Jill Spivey Caddell, Shinichi Nagata, and Ree Design Artwork.

Conflict of interest

Authors from Toyota Research Institute receive a small monetary stipend for publications. The remaining authors declare no conflict of interest.

Appendix A. Study tables

Below is the breakdown of all 69 studies spanning the various research methodologies discussed in this chapter. The tables include project identifiers, start dates of data collection, the country the study took place in, the total number of participants (P) and their age range, as well as the total number of hours put into the methodology (excluding data analysis) (**Tables A1–A3**).

Project	Date	US or JP	# of P	Age range	Total hours
In-depth interviews—ID					
ID-1	01/2019	US	25	65–78	15
ID-2	12/2018	US	22	60–65+	11
ID-3	11/2019	US	11	65–85+	11
ID-4	04/2020	US	8	59–96	8
ID-5	09/2020	US	41	50–65+	82
ID-6	06/2020	US	20	25+	1.6
ID-7	07/2020	JP	8	40–55	20
ID-8	11/2020	JP	12	25–39	15
ID-9	01/2021	JP	5	28–39	7.5
ID-10	03/2021	US	11	45+	11
ID-11	03/2021	JP	10	45+	10
ID-12	02/2021	US	10	65+	10
ID-13	02/2021	JP	10	65+	10
ID-14	09/2021	JP	10	65–76	10
ID-15	12/2021	JP	12	60–70	18
ID-16	12/2021	US	6	61–69	9
ID-17	03/2022	JP	10	22–73	10
ID-18	09/2022	JP	12	56–71	18
In-home ethnographies and walkthroughs—IHEW					
IHEW-1	03/2017	US	6	50–76	12
IHEW-2	01/2019	JP	12	45–71	30
IHEW-3	11/2022	JP	6	66–83	15

Findings in [33].

Table A1.
Study organization and key information grouped by research methodology.

Project	Date	US or JP	# of P	Age range	Total hours
Contextual inquiries—CI					
CI-1	03/2018	US	7	69–82	17.5
CI-2	09/2021	JP	10	65–76	10
Expert interviews—EI					
EI-1	08/2019	US	1	—	1
EI-2	08/2019	US	2	—	2
EI-3	04/2020	JP	2	—	2
EI-4	09/2021	JP	2	—	2
EI-5	08/2021	US	1	—	2.5
EI-6	03/2022	JP	2	—	2
EI-7	06/2022	US	12	—	26
EI-8	07/2022	US	8	—	41
EI-9	01/2023	JP	13	—	16
Focus groups—FG					
FG-1	03/2017	US	32	—	8
FG-2	01/2020	JP	28	35+	8
FG-3	06/2020	JP	27	35–79	22.5
FG-4	12/2020	JP	32	41–79	20
FG-5	08/2021	JP	52	26–85	24
FG-6	08/2021	US	12	26–80	8
FG-7	11/2022	JP	7	55–83	14
FG-8	06/2021	JP	15	55–79	6
FG-9	08/2022	JP	12	56–71	18
FG-10	09/2022	JP	15	54–82	6
FG-11	06/2023	JP	48	53–88	18
FG-12	06/2023	US	43	42–82	18
FG-13	08/2018	US	14	68–86	2
FG-14	05/2019	US	19	58–79	6
FG-15	05/2019	US	71	65–79	16.5
FG-16	07/2020	US	20	25+	1
Participatory design—PD					
PD-1	08/2023	JP	25	65–85	15
PD-2	09/2023	US	27	60–79	24
Design sprints—DSP					
DSP-1	12/2017	US	5	—	40
DSP-2	05/2018	US	7	—	21

Project	Date	US or JP	# of P	Age range	Total hours
DSP-3	06/2018	US	5	—	18.5
DSP-4	08/2018	US	4	68–77	4
Diary studies—DIARY					
DIARY-1	03/2018	US	15	—	11.4
DIARY-2	01/2022	JP	15	30–77	8.75
Surveys—SURVEY					
SURVEY-1	—	US	27	—	6.75
SURVEY-2	11/2016	US	97	—	32.3

Findings in [33].

Table A2.
Study organization and key information grouped by research methodology.

Project	Date	US or JP	# of P	Age range	Total hours
SURVEY-3	04/2019	US	1393	60–82	348.5
SURVEY-4	10/2019	US	20	30–71	6.6
SURVEY-5	07/2020	US	1364	—	363.7
SURVEY-6	06/2020	JP	80	25–65	26.6
SURVEY-7	05/2021	US	1001	20–65+	500.5
SURVEY-8	05/2021	JP	905	2–65+	452.5
SURVEY-9	05/2023	US	85	—	7
SURVEY-10	10/2023	US	98	32–77	8
Data set—DS					
DS-1	07/2020	JP	34	25–70	17
DS-2	03/2022	JP	32	19–75	16
DS-3	12/2022	US	10	—	25

Findings in [33].

Table A3.
Study organization and key information grouped by research methodology.

Author details

Katherine M. Tsui[1†*], Sarah Cohen[2†], Selma Sabanovic[3], Alex Alspach[1], Rune Baggett[4], David Crandall[3] and Steffi Paepcke[2]

1 Toyota Research Institute, Cambridge, MA, USA

2 Toyota Research Institute, Los Altos, CA, USA

3 Indiana University, Bloomington, IN, USA

4 Toyota Research Institute via HireArt, New York, NY, USA

*Address all correspondence to: kate.tsui@tri.global

†These authors contributed equally.

References

[1] Buehler R, Nobis C. Travel behavior in aging societies: Comparison of Germany and the United States. Transportation Research Record. 2010;**2182**(1):62-70. DOI: 10.3141/2182-09

[2] Lee R. The demographic transition: Three centuries of fundamental change. Journal of Economic Perspectives. 2002;**17**(4):167-190. DOI: 10.1257/089533003772034943

[3] Anton SD, Woods AJ, Ashizawa T, Barb D, Buford TW, Carter CS, et al. Successful aging: Advancing the science of physical independence in older adults. Ageing Research Reviews. 2015;**24**:304-327. DOI: 10.1016/j.arr.2015.09.005

[4] Piau A, Campo E, Rumeau P, Vellas B, Nourhashemi F. Aging society and gerontechnology: A solution for an independent living? The Journal of Nutrition, Health and Aging. 2014;**18**(1):97-112. DOI: 10.1007/s12603-013-0356-5

[5] Kirk D. Demographic transition theory. Population Studies. 1996;**50**(3):361-387. DOI: 10.1080/0032472031000149536

[6] Szreter S. The idea of demographic transition and the study of fertility change: A critical intellectual history. Population and Development Review. 1993;**19**(4):659-701. DOI: 10.2307/2938410

[7] Leeson GW. Increasing longevity and the new demography of death. International Journal of Population Research. 2014;**2014**(1):521523. DOI: 10.1155/2014/521523

[8] Vaupel JW, Villavicencio F, Bergeron-Boucher M-P. Demographic perspectives on the rise of longevity. Proceedings of the National Academy of Sciences. 2021;**118**(9):e2019536118. DOI: 10.1073/pnas.2019536118

[9] Okamoto Y. Health care for the elderly in Japan: Medicine and welfare in an aging society facing a crisis in long term care. BMJ British Medical Journal. 1992;**305**(6850):403. DOI: 10.1136/bmj.305.6850.403

[10] Broadbent E, Kerse N, Peri K, Robinson H, Jayawardena C, Kuo T, et al. Benefits and problems of health-care robots in aged care settings: A comparison trial. Australasian Journal on Ageing. 2016;**35**(1):23-29. DOI: 10.1111/ajag.12190

[11] Niemelä M, Van Aerschot L, Tammela A, Aaltonen I. A telepresence robot in residential care: Family increasingly present, personnel worried about privacy. In: Proceedings of the International Conference on Social Robotics (ICSR). Tsukuba, Japan: Springer; 2017. pp. 85-94. DOI: 10.1007/978-3-319-70022-9_9

[12] Baggett R, Simecek M, Tsui KM, Fraune MR. Temporal progression of four older adults through tech acceptance phases for a mobile telepresence robot in domestic environments. Robotics. 2024;**13**(7):95. DOI: 10.3390/robotics13070095

[13] Rheman JM, Baggett RP, Simecek M, Fraune MR, Tsui KM. Longitudinal study of mobile telepresence robots in older adults' homes: Uses, social connection, and comfort with tech. ACM Transactions on Human-Robot Interaction. 2024;**13**(4):1-41. DOI: 10.1145/3674956

[14] Smarr C-A, Mitzner TL, Beer JM, Prakash A, Chen TL, Kemp CC, et al. Domestic robots for older adults: Attitudes, preferences, and potential. International Journal of Social Robotics. 2014;**6**:229-247. DOI: 10.1007/s12369-013-0220-0

[15] Beuscher LM, Fan J, Sarkar N, Dietrich MS, Newhouse PA, Miller KF, et al. Socially assistive robots: Measuring older adults' perceptions. Journal of Gerontological Nursing. 2017;**43**(12):35-43. DOI: 10.3928/00989134-20170707-04

[16] Chen S-C, Jones C, Moyle W. Social robots for depression in older adults: A systematic review. Journal of Nursing Scholarship. 2018;**50**(6):612-622. DOI: 10.1111/jnu.12423

[17] Whelan S, Murphy K, Barrett E, Krusche C, Santorelli A, Casey D. Factors affecting the acceptability of social robots by older adults including people with dementia or cognitive impairment: A literature review. International Journal of Social Robotics. 2018;**10**(5):643-668. DOI: 10.1007/s12369-018-0471-x

[18] Padhan S, Mohapatra A, Ramasamy SK, Agrawal S. Artificial intelligence (ai) and robotics in elderly healthcare: Enabling independence and quality of life. Cureus. 2023;**15**(8):e42905. DOI: 10.7759%2Fcureus.42905

[19] Kemp CC, Edsinger A, Clever HM, Matulevich B. The design of stretch: A compact, lightweight mobile manipulator for indoor human environments. In: International Conference on Robotics and Automation (ICRA). Philadelphia, PA: IEEE; 2022. pp. 3150-3157. DOI: 10.1109/ICRA46639.2022.9811922

[20] Kamnik R, Bajd T. Standing-up robot: An assistive rehabilitative device for training and assessment. Journal of Medical Engineering and Technology. 2004;**28**(2):74-80. DOI: 10.1080/0309190032000112306

[21] Palacín J, Clotet E, Martínez D, Martínez D, Moreno J. Extending the application of an assistant personal robot as a walk-helper tool. Robotics. 2019;**8**(2):27. DOI: 10.3390/robotics8020027

[22] Kubota S, Nakata Y, Eguchi K, Kawamoto H, Kamibayashi K, Sakane M, et al. Feasibility of rehabilitation training with a newly developed wearable robot for patients with limited mobility. Archives of Physical Medicine and Rehabilitation. 2013;**94**(6):1080-1087. DOI: 10.1016/j.apmr.2012.12.020

[23] Lee S-H, Kim J, Lim B, Lee H-J, Kim Y-H. Exercise with a wearable hip-assist robot improved physical function and walking efficiency in older adults. Scientific Reports. 2023;**13**(1):7269. DOI: 10.1038/s41598-023-32335-8

[24] Lefeber N, Swinnen E, Kerckhofs E. The immediate effects of robot-assistance on energy consumption and cardiorespiratory load during walking compared to walking without robot-assistance: A systematic review. Disability and Rehabilitation: Assistive Tech. 2017;**12**(7):657-671. DOI: 10.1080/17483107.2016.1235620

[25] Sörensen S, Pinquart M. Developing a measure of older adults' preparation for future care needs. The International Journal of Aging and Human Development. 2001;**53**(2):137-165. DOI: 10.2190/1R0D-30TC-F4K1-F0DW

[26] Brach JS, VanSwearingen JM. Interventions to improve walking in older adults. Current Translational Geriatrics and Experimental Gerontology Reports. 2013;**2**:230-238. DOI: 10.1007/s13670-013-0059-0

[27] Woolhead G, Calnan M, Dieppe P, Tadd W. Dignity in older age: What do older people in the United Kingdom think? Age and Ageing. 2004;**33**(2):165-170. DOI: 10.1093/ageing/afh045

[28] Boldy D, Grenade L, Lewin G, Karol E, Burton E. Older people's decisions regarding 'ageing in place': A western australian case study. Australasian Journal on Ageing. 2011;**30**(3):136-142. DOI: 10.1111/j.1741-6612.2010.00469.x

[29] Eckert JK, Morgan LA, Swamy N. Preferences for receipt of care among community-dwelling adults. Journal of Aging and Social Policy. 2004;**16**(2):49-65. DOI: 10.1300/J031v16n02 04

[30] Matsumoto H, Naruse T, Sakai M, Nagata S. Who prefers to age in place? Cross-sectional survey of middle-aged people in Japan. Geriatrics and Gerontology International. 2016;**16**(5):631-637. DOI: 10.1111/ggi.12503

[31] Wang S, Bolling K, Mao W, Reichstadt J, Jeste D, Kim H-C, et al. Tech to support aging in place: Older adults' perspectives. In: Healthcare. Vol. 7. MDPI; 2019. p. 60. DOI: 10.3390/healthcare7020060

[32] Wiles JL, Leibing A, Guberman N, Reeve J, Allen RE. The meaning of "aging in place" to older people. The Gerontologist. 2012;**52**(3):357-366. DOI: 10.1093/geront/gnr098

[33] Tsui K, Baggett R. Uncovering Older Adult and in-Home Needs: Applying User-Centered Research Methodologies to Inform Robotics Development and a Call to Action, 2024. Available from: https://www.osf.io/8729k

[34] Knapp J, Zeratsky J, Kowitz B. Sprint: How to Solve Big Problems and Test New Ideas in Just Five Days. Simon and Schuster; 2016. ISBN 978-1-5011-2174-6

[35] Toyota Research Institute. Available from: https://www.tri.global

[36] Well Spousal Assocation. About Spousal Caregiving, 2023. Available from: https://wellspouse.org/who-we-are/statistics.html.

[37] Danely J. Hope in an ageing Japan: Transience and transcendence. Contemporary Japan. 2016;**28**(1):13-31. DOI: 10.1515/cj-2016-0002

[38] Ishizuka-Inoue M, Kawaguchi A, Kashima S, Nagai-Tanima M, Aoyama T. Differences in physical activity and mental function according to the employment status of elderly Japanese. Journal of Occupational Health. 2023;**65**(1):e12411. DOI: 10.1002/1348-9585.12411

[39] Tan M-E, Sagayadevan V, Abdin E, Picco L, Vaingankar J, Chong SA, et al. Employment status among the Singapore elderly and its correlates. Psychogeriatrics. 2017;**17**(3):155-163. DOI: 10.1111/psyg.12206

[40] Clarke V, Braun V. Thematic analysis. The Journal of Positive Psychology. 2017;**12**(3):297-298. DOI: 10.1080/17439760.2016.1262613

[41] Carmeli E, Patish H, Coleman R. The aging hand. The Journals of Gerontology Series A: Biological Sciences and Medical Sciences. 2003;**58**(2):M146-M152. DOI: 10.1093/gerona/58.2.M146

[42] Eckstrom E, Neukam S, Kalin L, Wright J. Physical activity and healthy aging. Clinics in Geriatric Medicine. 2020;**36**(4):671-683

[43] Frank JS, Patla AE. Balance and mobility challenges in older

adults: Implications for preserving community mobility. American Journal of Preventive Medicine. 2003;**25**(3):157-163. DOI: 10.1016/S0749-3797(03)00179-X

[44] Liverman CT, Yaffe K, Blazer DG. Cognitive Aging: Progress in Understanding and Opportunities for Action. National Academies Press; 2015

[45] Kakara R. Nonfatal and fatal falls among adults aged 65 years—United States, 2020-2021. MMWR. Morbidity and Mortality Weekly Report. 2023;**72**:938-943. DOI: 10.15585/mmwr.mm7235a1

[46] AHS stairs. American Housing Survey (AHS)—Table Creator. 2023. Available from: https://www.census.gov/programs-surveys/ahs/data/interactive/ahstablecreator.html?s_areas=00000&s_year=2021&s_tablename=TABLE1&s_bygroup1=1&s_bygroup2=1&s_filtergroup1=1&s_filtergroup2=1

[47] Pereira SG, C. B. d. Santos, M. Doring, and M. R. Portella. Prevalence of household falls in long-lived adults and association with extrinsic factors. Revista Latino-Americana de Enfermagem. 2017;**25**:e2900. DOI: 10.1590/1518-8345.1646.2900

[48] Edwards N, Dulai J, Rahman A. A scoping review of epidemiological, ergonomic, and longitudinal cohort studies examining the links between stair and bathroom falls and the built environment. International Journal of Environmental Research and Public Health. 2019;**16**(9):1598. DOI: 10.3390/ijerph16091598

[49] Sophonratanapokin B, Sawangdee Y, Soonthorndhada K. Effect of the living environment on falls among the elderly in Thailand. Southeast Asian Journal of Tropical Medicine and Public Health. 2012;**43**(6):1537. ISSN 0125-1562

[50] Pu-Lin Y, Zhao-Hui Q, Jing S, Zhang J, Mei-Zhe X, Zheng-Lai W, et al. Prevalence and related factors of falls among the elderly in an urban community of Beijing. Biomedical and Environmental Sciences. 2009;**22**(3):179-187. DOI: 10.1016/S0895-3988(09)60043-X

[51] Komisar V, van Schooten KS, Aguiar OM, Shishov N, Robinovitch SN. Circumstances of falls during sit-to-stand transfers in older people: A cohort study of video-captured falls in long-term care. Archives of Physical Medicine and Rehabilitation. 2023;**104**(4):533-540. DOI: 10.1016/j.apmr.2022.10.012

[52] Sagari A, Tabira T, Maruta M, Tanaka K, Iso N, Okabe T, et al. Risk factors for nursing home admission among older adults: Analysis of basic movements and activities of daily living. PLoS One. 2023;**18**(1):e0279312. DOI: 10.1371/journal.pone.0279312

[53] Abidi H, Cianchetti M. On intrinsic safety of soft robots. Frontiers in Robotics and AI. 2017;**4**. DOI: 10.3389/frobt.2017.00005. . ISSN 2296-9144. Available from: https://www.frontiersin.org/journals/robotics-and-ai/articles/10.3389/frobt.2017.00005

[54] Alspach A, Beaulieu A, Tsui KM. Meet Punyo, TRI's Soft Robot for Whole-Body Manipulation Research, 2024. Available from: https://medium.com/toyotaresearch/meet-punyo-tris-soft-robot-for-whole//-body-manipulation-research-//949c934ac3d8.

[55] Zhang M, Barreiros J, Onol AO. Plan-Guided Reinforcement Learning for Whole-Body Manipulation, 2023.

Available from: https://arxiv.org/abs/2310.12263.

[56] Hsu L-J, Stafford PB, Khoo W, Swaminathan M, Amon KJ, Sato H, et al. "Give it time:" longitudinal panels scaffold older adults' learning and robot co-design. In: Proceedings of the 2024 ACM/IEEE International Conference on Human-Robot Interaction (HRI '24). New York, NY: Association for Computing Machinery; 2024. pp. 283-292

[57] Kamino W, Hsu L-J, Joshi S, Randall N, Agnihotri A, Tsui KM, et al. Making meaning together: co-designing a social robot for older adults with Ikigai experts. International Journal of Social Robotics. 2023;**15**(6):983-998

[58] Randall N, Joshi S, Kamino W, Hsu L-J, Agnihotri A, Li G, et al. Finding ikigai: How robots can support meaning in later life. Frontiers in Robotics and AI. 2022;**9**:1011327

[59] G. I. F. Health and P. of Life Development Foundation. Toward a New Stage in the Age of 100 Years of Life. 2019.

[60] Hasegawa A, Fujiwara Y, Hoshi T. Trends in IKIGAI research in Japan from 2000 to 2014: The "renaissance" of Ikigai research. Healthy Active Aging. 2015;**21**(2015):60-143

[61] LuxAI. Available from: https://luxai.com

[62] Hsu L-J, Kamino W, Khoo W, Tsui K, Crandall D, Šabanović S. Working together toward ikigai: Co-designing robots that can help us achieve meaning and purpose in life. XRDS: Crossroads, The ACM Magazine for Students. 2023;**30**(1):38-45

[63] Hsu L-J, Khoo W, Swaminathan M, Amon KJ, Muralidharan R, Sato H, et al. Let's talk about you: Development and evaluation of an autonomous robot to support Ikigai reflection in older adults. In: IEEE International Conference on Robot and Human Interactive Communication (RO-MAN). 2024. Available from: https://api.semanticscholar.org/CorpusID:270367109

Chapter 2

Use of Senior Technology Acceptance Model (STAM) for Social Robots Studies

Kelvin Cheng Kian Tan

Abstract

This chapter explores the adaptation of technology acceptance model (TAM) to senior technology acceptance model (STAM). The extension provides more focus on the perception and acceptance of technology by older persons for researchers to have a better understanding of the multidimensional facets of ageing needs. In the studies conducted in Hong Kong and Singapore, we have used the assessment of two different social robots. For the Hong Kong study, Humanoid social robot, Kabochan, was deployed as an intervention in seven nursing homes to understand the acceptance by residents living with dementia, on the other hand, in a separate joint study between Hong Kong and Singapore, a Japanese affectionate robot, LOVOT, was chosen. Both studies used STAM and a number of social well-being measures to assess the benefits of social robots in the health and care of older persons. STAM has demonstrated its usefulness and relevance in building evidence and correlations to elucidate the varied research objectives. Moving forward, as gerontechnology is populated with advancement in wearables, Internet of Things (IoT), artificial intelligence (AI), and robotics, STAM is likely have a stronger place in the research and evaluation scene. The collection of evidence is to appreciate the complexity of factors in the adoption of gerontechnology.

Keywords: STAM, older persons, social robot, wellbeing, quality of life, AI

1. Introduction

1.1 Ageing society needs more technologies

Globally, many countries are facing an ageing population that is escalating at an alarming rate. In 2023, the United Nations (UN) reported that the world's population of people aged 65 years and older would be more than doubled to 1.6 billion in 2050. The rate of rising number of those people in the age group of 80 years and older is accelerating. Decreasing fertility rates and improving life expectancy have led to this demographic phenomenon. The Total Fertility Rate (TFR) has been on the decline in all Southeast Asian countries. Singapore had a TFR of 1.2 in 2014. Many women stay childless in Singapore. It is inevitable that Singapore will enter the status of super-ageing nation by 2030 [1]. The family structure will be affected by these changes and older persons will likely be more isolated.

Ageing adults could encounter physical and cognitive decline, which affects their Activities of Daily Living (ADLs), and therefore there is the need for more carers to meet the new needs of this social development. However, the health and care systems are strained by the availability of manpower to support an ageing population, and there is an overall decline in caregiver-to-senior ratio. An example of the health and care sector stress is demonstrated by the COVID-19 pandemic, when Singapore healthcare workers experienced stress and job burnout [2] from the uptake in workload.

In the community, loneliness and social isolation is another growing issue, particularly among older persons particularly in the new demographic shift in many societies. Loneliness is a state of distress and unpleasantness between the desire for actual and desired social connections. Social isolation is the lack of social network [3]. Older persons can still experience loneliness in a crowd. Besides, there are more older persons who are staying alone with limited social contacts and engagements. Loneliness is classified as a subjective emotional state, whereas social isolation is an objectively quantifiable variable. Studies have shown that there is measurable influence attributed to both objective and subjective social isolation on decreased mortality [4]. Post COVID-19, more emphasis has been placed on the topic of social well-being of older persons and the impact on their family members. During the challenge of pandemic, it has disproportionately affected people living with dementia and their carers [5]. A study has showed that loneliness is a main driver of social well-being [6]. It is imperative to nip the problem in the bud before the feeling of social isolation and loneliness can potentially spiral into more health issues.

In general, technologies can be the best alternative to provide assistance in managing the problems and challenges posed by ageing as the older people could have a better quality of life, more independence and be socially engaged [7]. More studies were conducted to illuminate the deployment of technology as an abler for fostering communication and engagement between older persons and the communities. In the era of social media, IoT, wearables, robots and smart home solutions, technology can facilitate the delivery of health and care to manage loneliness and social isolation. More assessments on the use of technology can generate results from these studies, which can be highly relevant globally in tackling the prejudices of technology in overcoming the issues faced by old people [4].

In the field of gerontechnology, different developments and applications of technology for older persons are explored and promoted. A plethora of mobile applications and information and communication technology (ICT) solutions were launched to help older persons and their families to connect more efficiently with each other. Means of accessing resources in the communities of health and care, and thereby improving their mental and physical well-being are prevalent [8]. In addition, studies showed that peer programme inviting older persons to participate in activities can reduce loneliness, depression and low entry to socialising [9]. Thus, a technology designed to match people with similar backgrounds and interests is becoming more popular for older persons and their peers.

More recent research has discovered that social robots that support the older population in the areas of "mobility, self-care, interpersonal interaction" resulted in profound positive effects. Social robots are designed to bring companionship and partnership to older persons and their caregivers.

A popular example is PARO (a seal robot designed in Japan) as a therapeutic pet for older persons living with dementia. It has been used in care settings for more than 10 years in many countries. Compared with living animals, social robots are more suitable in care settings with older persons. Among the benefits are lesser care and safer use. Studies with PARO have found to improve anxiety and psychological symptoms of people living with dementia [10].

In the research study in Singapore and Hong Kong, a popular social robot, LOVOT, invented in Japan was selected. LOVOT was created by Groove X in 2019 with the specific goal of being a companion to families. LOVOT is huggable and has warm features that include internal temperature regulation, which allows it to keep an average human body temperature, and the ability to interact intuitively with humans. Its lively pair of eyes, chirpy tone and unique name can be customised through a smartphone application to create more personalisation. A team of designers comprising of professional animators with deep experience in animation and game industry created the award-winning design that has captivated many with its cuteness and attractive User Interface. Multiple trials and experiments of the expressions and movements conducted were critical to its eventual appealing nature. Kaname Hayashi who is the creator of LOVOT described the core concepts that the product embraces: simple shape, horn with multiple sensors and a soft, warm body [11]. Sensors embedded in LOVOT's body and a camera at the top of its head recognise the individual human and sounds.

2. Perception of technology by older persons

Older persons are overwhelmed by a torrent of technologies and have shown the development of positive attitude towards it. However, their adoption of technology and interest is unlikely to be as high as younger people [12]. Their behaviours vary, especially in communication, customer service, healthcare and home-based services. In order to comprehend and predict the older persons' use of technology, it is essential to determine the factors that affect their acceptance and use of technology.

The two models: technology acceptance model (TAM) and unified theory of acceptance and use of technology (UTAUT) have been used in a number of studies to understand the acceptance of technology in general. In TAM, perceived usefulness (PU) and perceived ease of use (PEOU) are key attitudinal factors in explaining the acceptance and use of ICT solutions. PU was described as "the degree to which a person believes that using the particular technology would enhance his/her job performance". PEOU was defined as "the extent to which a person believes that using a technology is free of effort" [13]. Combined, PU and PEOU can determine the user's attitude towards technology. PU and AT in turn will determine the behavioural intention (BI) to use the technology introduced, which affects the actual utilisation.

Venkatesh et al. [14] formulated UTAUT that identified five direct determinants and four moderators to increase the predictive power of the model.

Though both TAM and UTAUT have been used widely and analysed [15], the needs of older persons are different due to the heterogeneity, which becomes more pronounced with age [16]. Studies have shown that person-centred design is needed to increase the adoption of products serving the potentials with no constraints. The dimensions of physical, social and psychological can help to understand the users' interactions with technology and products [17].

3. Senior technology acceptance model

With the knowledge of TAM and UTAUT models and adding the focus to older persons, senior technology acceptance model (STAM) was developed. It is an extension of the previous TAM versions and theories to include the age-related health and abilities of an older person. The unique needs of the older persons based on their psychological and sociology dimensions will help to shape and create better and more robust products, which can better respond to their needs. Some of the determinants are User Interface Design, Navigation of Apps, Training and familiarity with technology, feedback of their experience, consistency in design and terminology, flexibility and customisation, accessibility and even documentation.

STAM was evolved from a study conducted in Hong Kong to test the model, with the primary focus of evaluating the acceptance of gerontechnology among 1012 elderly Chinese who were aged 55 and over. A total of 11 constructs comprising of 38 items with multidimensional measure of older person's acceptance of technology were included in the study. The study demonstrated that the heterogeneity of older persons and their attributes, comprising of age, gender, education level, gerontechnology self-efficacy and anxiety, and health and ability characteristics, together with the facilitating conditions (FC), are explicitly and directly linked to technology acceptance. Beyond the age-related attributes, STAM has identified additional factors that could contribute to the technology acceptance of older Hong Kong Chinese [18]. Physiological and psychological abilities can be affected with age, which in turn will affect the acceptance of technology [19]. The benefits of using STAM as a measure to understand the application of gerontechnology can assist in overcoming the problems and challenges to build more useful products for older persons to lead better lives (**Figure 1**) [20, 21].

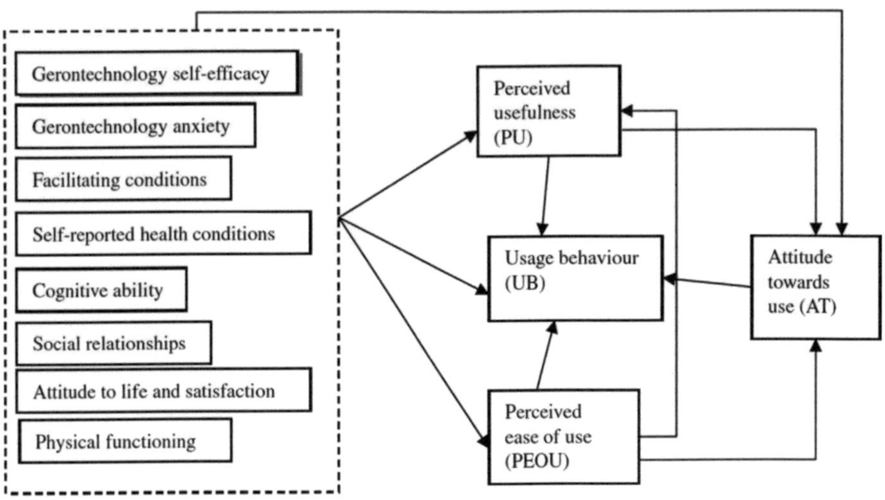

Figure 1.
Senior technology acceptance model (STAM).

4. STAM-14: A shorter version of STAM

Further use of 38-item STAM in research showed that older adults with lower education encountered difficulty filling in the questionnaires independently, and it will take more than 30 minutes to administer the interview. This can limit its use due to the required time to complete the questionnaire and also affect its utilisation for rapid use as assessment.

A shorter version of the questionnaire, which is validated and reliable, is likely to improve the usage of the instrument. Through a sequential item-reduction strategy combined with convergent and discriminant validity analysis using confirmatory factor analysis (CFA) and review of resultant terms by experts, STAM-14 was born. A brief version that had a 4-factor structure comprising of technology acceptance constructs and age-related health factors was designed. In research that is constrained by administrative time and older persons' availability, STAM-14 can be used, as it aimed to reduce the respondent burden and optimise its utilisation (**Table 1**) [21].

5. Study 1: The Hong Kong study using STAM to measure the changes of technology acceptance among older people with dementia with social robots engagement

Studies conducted in United States, Australia and the United Kingdom have shown that at least 50% of their residents in long-term care facilities are living with dementia. The staff will need to cope with the more demanding behaviour of these residents [22].

Animal-assisted therapy (AAT) can reduce aggression and agitation of people living with dementia and also promote their social behaviour [23]. However, it is not easy to facilitate visits. With the introduction of more social robots into the gerontology scene, studies have shown that pet robot is expected to yield positive outcome for people living with dementia including concerning the quality of life [24]. So far, there is limited understanding of changes in technology acceptance, lack of direct exposure to technology and rigorous study design. This Hong Kong study examined the change in technology acceptance after residents with dementia in the long-term care facilities nursing homes have direct interaction with a social robot, Kabochan.

First introduced in 2011, the Japanese social robot is a 3-year-old boy look-alike and is priced at an affordable cost of USD$ 300. It is of 28 cm height and weighs only 680 g. It has a collection of 13 songs and possesses 450 phrases and words. There are five sensors embedded in it to improve its responsiveness and interactivity. It is sensitive to various stimulations from audio, light and movements. During verbal

Short STAM subscales	**No. of items**	**Cronbach's α**	**Composite reliability**	**Average variance extracted**
Attitudinal beliefs	3	0.915	0.921	0.795
Control beliefs	4	0.846	0.820	0.534
Gerontology anxiety	2	0.847	0.850	0.793
Health	5	0.817	0.805	0.455

Table 1.
Cronbach's alpha, composite reliability, and average variance extraction of short STAM.

engagement, it will nod its head and thank anyone who pats its head. The range of expressions will become more personalised if the owner spends more time with it. It is designed with the intent of reducing loneliness experienced by the people living with dementia as a friendly companion. Kabochan possesses pre-programmed speaking words to address its owner including "Grandma" and "Grandpa" in Japanese. Moreover, the Kabochan offers a few physical exercise modes, which include a posing game, raising the flag game, as well as vocal singing exercises (**Figure 2**) [25].

The technology acceptance was assessed using Randomised Control ABAB withdrawal design for a period of 32 weeks. The participants were made up of 103 residents clinically diagnosed with dementia, with a mean age of 87.2 years. They were recruited from seven long-term care facilities in Hong Kong and were randomly allocated to either Kabochan engagement group or control group. Participants in the engagement group interacted with Kabochan in an individual, non-facilitated approach. The behavioural interactions with Kabochan were observed by frontline care workers and recorded to understand the utilisation level as a form of user behaviour. Questionnaire surveys were taken placed at pre- and post-exposure with Kabochan to measure attitudes and beliefs towards technology. The STAM questionnaire was based on the main constructs of technology acceptance: Attitudes towards Technology (AT), Perceived Usefulness (PU), Perceived Ease of Use (PEOU), Technology Self-Efficacy (SE), Technology Anxiety (AT) and Facilitating Conditions (FC).

Outcome measures were introduced in the study to relate the older persons' social and psychological well-being to their acceptance of social robots (**Figure 3**).

Using mixed multivariate analysis of variance (MANOVA), there is a significant group-by-time interaction effect at week 32. The PEOU has small-to-moderate effect size ($F = 4.239$, $p = 0.042$, $\eta\rho^2 = 0.043$), with controlling of covariates at baseline. It showed that there are significant improvements in PEOU for the intervention group compared with the control group at week 32 (**Table 2**). However, there is no evidence pertaining to AT, PU, FC, ANX and SE after the use of intervention with Kabochan social robot at week 32. No significant covariate-by-time intervention effect was

Figure 2.
Social humanoid robot: Kabochan.

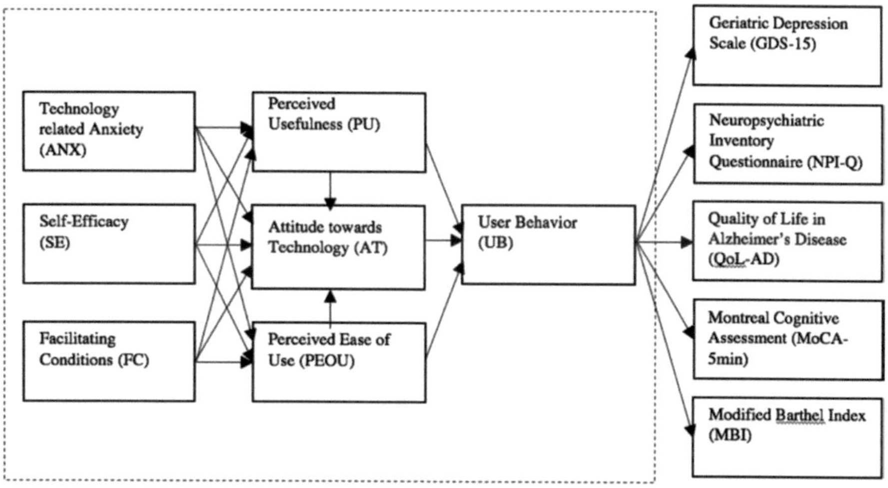

Figure 3.
STAM and outcome measures.

Outcome measures	Week 1		Week 32		Group-by-time interaction effects		
	Mean	SD	Mean	SD	F	p	η_p^2
Attitudes towards technology							
Usual care	11.45	(5.24)	11.65	(5.07)	1.649	0.202	0.017
Engagement	10.50	(5.16)	11.87	(4.81)			
Perceived usefulness							
Usual care	15.08	(8.03)	16.00	(7.82)	1.569	0.213	0.016
Engagement	13.50	(7.43)	15.98	(7.07)			
Perceived ease of use							
Usual care	8.51	(5.75)	7.00	(4.63)	4.239	0.042	0.043
Engagement	7.50	(5.76)	8.42	(4.48)			
Facilitating conditions							
Usual care	23.00	(11.03)	21.51	(9.85)	1.684	0.198	0.017
Engagement	22.63	(9.86)	23.85	(8.13)			
Technology self-efficacy							
Usual care	8.10	(5.01)	8.18	(4.79)	1.248	0.267	0.013
Engagement	7.31	(5.27)	8.73	(4.47)			
Technology anxiety							
Usual care	8.24	(4.81)	12.57	(7.36)	2.987	0.087	0.030
Engagement	8.94	(5.05)	11.27	(4.62)			

Note:
η_p^2: Partial eta-square.
Covariates appearing in the model are evaluated at the following values: age = 87.14, Activity of Daily Living at baseline = 44.3, Montreal Cognitive Assessment 5-minute Protocol at baseline = 5.6, Geriatric Depression Scale at baseline = 6.5 and Neuropsychiatric Inventory-Questionnaire at baseline = 3.2.

Table 2.
Changes in technology acceptance between usual care and social robot engagement groups.

found in any of the STAM outcomes ($p > 0.05$). Similarly, there is no significant covariate-by-time interactions effect in any of the STAM outcomes ($p > 0.05$). As a result, the hypothesis of a significant improvement in technology acceptance for Kabochan engagement as an intervention was partially supported [26].

The study clocked 17,248 engagement observations for social robot intervention group, 63.93% was non-engagement. Among the engagement activities with Kabochan: cuddling ranked the highest (13.63%), followed by talking (8.66%), coaxing to sleep (4.68%), moving feet and arms (3.53%), carrying (2.42%), tidying appearance (1.52%), tickling foot (1.41%) and feeding (0.22%).

In addition, the results showed that there were no significant between-group differences in any of the STAM measures ($p > 0.05$) when the engagement level was included for analysis as a covariate. However, the level of behavioural engagement and magnitude of change in attitude towards technology ($F = 10.17$, $p = 0.01$, $\eta\rho^2 = 0.10$) and effect size was moderate to large. Similarly, the level of engagement significantly affected the changes in perceived usefulness with a moderate effect size ($F = 6.19$, $p = 0.01$, $\eta\rho^2 = 0.06$).

The main findings showed that using the STAM questionnaire to assess the exposure of an older person living with dementia to social robot has clearly demonstrated the benefits in changing perceived ease of use (PEOU) in long-term care setting. Therefore, the use of direct engagement with a social robot can potentially further improve PEOU of older persons living with dementia [26]. Besides, the social robots can relieve caregiving stress while reducing the older residents' problematic behaviours. Through a balanced complement of social robots with the mainstream human touch of the carers, the long-term care setting can be a conducive place with a high tech and high touch ambience for the older residents, the carers and even family members in the active ageing community.

6. Study 2: Acceptance of social robot among singleton older adults in Singapore

In the backdrop of impending super-ageing societies, there is a rising number of singleton households in Singapore. In particular, older adult singletons are likely to face more psychosocial challenges, such as the lack of companionship and loss of self-control [27]. Such a decline in mental well-being and health of an older person can impact their physical health. In the disruptive COVID-19 period, the measures taken to control the pandemic have resulted in negative implications on the social isolation and loneliness of older people. Therefore, the gerontology researchers and scholars were concerned about the "loneliness epidemic" spreading among the ageing community [28]. Due to a paucity of literature on the efficacy of assistive technology in this area, a study was launched to explore such use of assistive technology in the form of a Japanese social robot, LOVOT, as companion for singletons in the community (**Figure 4**).

The study conducted a baseline assessment with the STAM-14 tool followed by three 15-minute interactive sessions with a LOVOT over a period of 4 weeks. A mixed-methods design was used to measure the acceptance and quality of interaction between LOVOT and the single older persons. At baseline assessment, the participants were interviewed using STAM-14 [21], Older People's Quality of Life (OPQL) [29], Loneliness (L) [30], Subjective Happiness (SH) [31], Cultural inclinations [32], Willingness to pay and LOVOT's perceived sociability [32] and system usability (**Figure 5**) [33].

Figure 4.
Social robot: LOVOT.

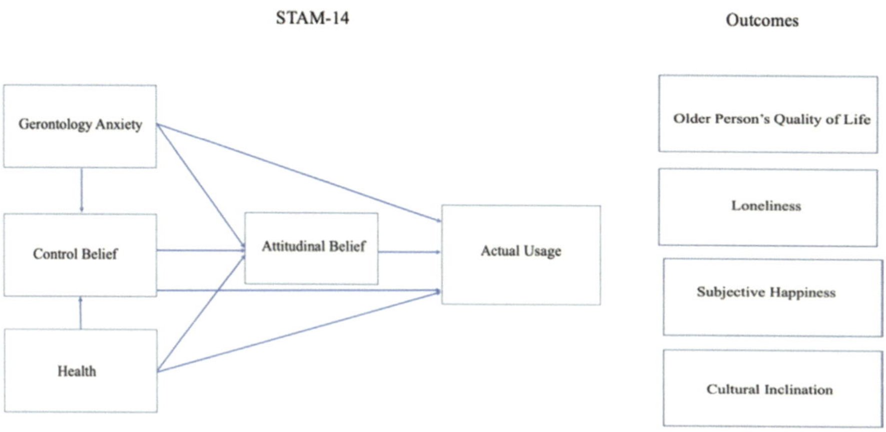

Figure 5.
STAM-14 and outcome measures.

A total of 15 participants, 60–75 years old, with no known cognitive or mental issues, were recruited in Singapore. The quantitative measures adopted in the study and the observational data from the interaction sessions during pre- and post-interaction interviews were thoroughly analysed to investigate the factors affecting the acceptance of LOVOT among older singletons in Singapore.

Prior to the first interaction with LOVOT, the STAM-14 tool was used to ask participants about their use and acceptance of technology. In the context of technology, participants were asked the type of device they used most of the time, their perception of system usability of the technology and the resources they could seek for advice. In terms of the difficulties encountered with technology, the participants were asked questions that assessed their degree of acceptance and anxiety. After the last interaction with LOVOT, the participants were asked questions that expressed their emotional and experiential feedback after playing with LOVOT in comparison of living pets and their acceptance. All questions were structured objectively to avoid any response biases. In conducting the analysis of the responses, more could be gleaned on the acceptance of LOVOT by singletons. The team made an audio recording of the interviews for transcribing purposes during post-interview analysis. In total, three categories of observations were supported: Verbal actions, Physical actions and Combined actions (simultaneous verbal and physical actions). The researchers shall infer the common patterns and contents from these observations.

Finding of the correlation analysis of the five measures of STAM showed several significant results for each time point (**Tables 3–6**). At To, T1 and T3, attitudinal belief (AB) and control belief (CB), AB and behavioural intention (BI) are significantly correlated. At T0, there was a significant positive correlation between CB and health conditions (HC). However, at T3, there was a significant negative correlation between AB and HC.

Social robot LOVOT is well received as it created a sense of meaning for the participants and so the positive correlation of AB with the other constructs of STAM is in line with the hypothesis of technology acceptance and role of social robot as companion. Therefore, the positive correlations between the subscales (CB, HC, AB and BI) within the STAM model indicated that the older persons' responses were consistent across the subscales (**Table 7**).

Upon completion of the exploratory study, a set of interventional protocol developed in both English and traditional Chinese was made available. The incorporation of STAM-14 and social well-being measures will provide guidance for use of social robot, LOVOT, in intervention studies for individual and group in community or institutional settings.

The finding from this study serves to inform practitioners, carers and policymakers on the possible integration of any forms of social robots into the care model of older singletons in the community. Future studies are lined up as LOVOT has started to be commercially available in places outside of Japan.

7. What does the future hold for STAM?

More countries around the world are entering the era of super-ageing status. In the advent of new generation of gerontechnology using Virtual Reality (VR), Metaverse, Artificial Intelligence (AI), Generative AI, Wearables, Robots and Telemedicine, evidence-based studies will be highly demanded. More collaboration between technology providers, healthcare providers and older persons remains vital in promoting inclusive, person-centred and ethically complied technological products and services.

It is evident that technology adoption by older persons is affected by a range of determinants. These are gender, cost, education and technology literacy. A deeper understanding and insight into use of technology with the approach of the person-centred approach will create more successes for the industry.

Arguably, STAM is going to be used more often as an assessment tool in measuring technology acceptance and adoption by an older person. Complementing the empirical

	T0: Baseline assessment	T1: First interaction (15 minutes)	T2: Second interaction (15 minutes)	T3: Third interaction (15 minutes)
Survey outcomes measured	• STAM[a] • Sociability of robot • Loneliness • Older people's quality of life and general quality of life • Subjective happiness • System usability • Willingness to pay • Demographics • Cultural values	• STAM • Sociability of robot • Loneliness • Older people's quality of life and general quality of life • Subjective happiness • System usability • Willingness to pay	• Loneliness • Older people's quality of life and general quality of life • Subjective happiness	• STAM • Sociability of robot • Loneliness • Older people's quality of life and general quality of life • Subjective happiness • System usability • Willingness to pay
Interviews	N/A[b]	Pre- and post-interview	N/A	Post-interview

[a]STAM: Senior technology acceptance model.
[b]N/A: Not applicable.

Table 3.
Survey and interview outcomes breakdown.

	AB	CB	GA	HC	BI	SR	L	OPQL	SH
AB	1.00	0.57	0.09	0.45	0.76	0.56	−0.21	0.15	0.43
CB	0.57	1.00	−0.24	0.60	0.67	0.25	−0.56	0.43	0.68
GA	0.09	−0.24	1.00	0.09	0.04	0.16	0.31	−0.31	−0.33
HC	0.45	0.60	0.09	1.00	0.09	−0.10	−0.75	0.53	0.65
BI	0.76	0.67	0.04	0.09	1.00	0.72	−0.17	0.15	0.32
SR	0.56	0.25	0.16	−0.10	0.72	1.00	0.11	−0.06	0.19
L	−0.21	−0.56	0.31	−0.75	−0.17	0.11	1.00	−0.59	−0.49
OPQL	0.15	0.43	−0.31	0.53	0.15	−0.06	−0.59	1.00	−0.59
SH	0.43	0.68	−0.33	0.65	0.32	0.19	−0.49	−0.59	1.00

AB = attitudinal beliefs, CB = control beliefs, GA = gerontechnology anxiety, HC = health conditions, BI = behavioural intention, SR = sociability of robot, L = loneliness, OPQL = older people's quality of life, SH = subjective happiness.

Table 4.
Correlation between measures (T0).

	AB	CB	GA	HC	BI	SR	L	OPQL	SH
AB	1.00	0.76	0.17	0.33	0.91	0.82	0.16	0.20	0.03
CB	0.76	1.00	0.39	0.30	0.77	0.60	0.41	0.33	−0.09
GA	0.17	0.39	1.00	0.21	0.26	0.19	0.15	0.48	−0.14
HC	0.33	0.30	0.21	1.00	0.21	0.44	0.11	0.06	0.27
BI	0.91	0.77	0.0426	0.21	1.00	0.87	0.25	0.28	−0.07
SR	0.82	0.60	0.19	0.44	0.87	1.00	0.24	0.16	−0.07
L	0.16	0.41	0.15	0.11	0.25	1.24	1.00	−0.13	−0.20
OPQL	0.20	0.33	0.48	0.06	0.28	0.16	−0.13	1.00	0.44
SH	0.03	−0.09	−0.14	0.27	−0.07	−0.07	−0.20	0.44	1.00

AB = attitudinal beliefs, CB = control beliefs, GA = gerontechnology anxiety, HC = health conditions, BI = behavioural intention, SR = sociability of robot, L = loneliness, OPQL = older people's quality of life, SH = subjective happiness.

Table 5.
Correlations between measures (T1).

	L	OPQL	SH
L	1.00	−0.39	−0.39
QL	−0.39	1.00	0.84
SH	−0.39	0.84	1.00

L = loneliness, OPQL = older people's quality of life, SH = subjective happiness.

Table 6.
Correlations between measures (T2).

analysis from quantitative data collected by STAM and other social well-being measures will provide a more holistic assessment of the older persons and gradually nudge them closer toward improving their attitude of technology. Continuous sharing of the best practices that overcome the impediments can strive for an empathic and more inclusive community empowered by the technology innovations.

	AB	CB	GA	HC	BI	SR	L	OPQL	SH
AB	1.00	0.73	−0.17	0.13	0.93	0.85	0.40	−0.21	0.09
CB	0.763	1.00	−0.13	0.47	0.85	0.59	0.42	0.13	0.20
GA	−0.17	−0.13	1.00	0.17	−0.15	0.08	0.10	0.01	−0.27
HC	0.13	0.47	0.17	1.00	0.17	−0.13	−0.11	0.76	0.69
BI	0.93	0.85	−0.15	0.17	1.00	0.85	0.28	−0.05	0.18
SR	0.85	0.50	−0.08	−0.13	0.85	1.00	0.26	−0.36	−0.05
L	0.40	0.42	0.10	−0.11	0.28	0.26	1.00	−0.30	−0.70
OPQL	−0.21	0.13	0.02	0.76	−0.05	−0.36	−0.30	1.00	0.75
SH	0.09	0.20	−0.27	0.69	0.18	−0.05	−0.27	0.75	1.00

AB = attitudinal beliefs, CB = control beliefs, GA = gerontechnology anxiety, HC = health conditions, BI = behavioural intention, SR = sociability of robot, L = loneliness, OPQL = older people's quality of life, SH = subjective happiness.

Table 7.
Correlations between measures (T3).

Acknowledgements

Hong Kong Sheng Kung Hui Welfare Council.
Orange Valley Active Ageing Center, Singapore.
Ministry of Education.
Nunchi Marine Singapore.
Sao Po Center on Ageing, The University of Hong Kong.

Author details

Kelvin Cheng Kian Tan
Singapore University of Social Sciences, Singapore

*Address all correspondence to: kelvintanck@suss.edu.sg

References

[1] Yeung W-JJ. Demographic and Family Transition in Southeast Asia. Springer International Publishing AG; 2022. pp. 110-111

[2] Teo I, Chay J, Cheung YB, Sung SC, Tewani KG, Yeo LF, et al. Healthcare worker stress, anxiety and burnout during the COVID-19 pandemic in Singapore: A 6-month multi-centre prospective study. PLoS One. 2021;**16**(10):e0258866-e0258866. DOI: 10.1371/journal.pone.0258866

[3] Pronko NH. In: Peplau LA, Perlman D, editors. Loneliness: A Sourcebook of Current Theory, Research and Therapy. Vol. 33. Granville, Ohio: Denison University; 1983. p. 147

[4] Holt-Lunstad J, Smith TB, Baker M, Harris T, Stephenson D. Loneliness and social isolation as risk factors for mortality: A meta-analytic review. Perspectives on Psychological Science. 2015;**10**(2):227-237. DOI: 10.1177/1745691614568352

[5] West E, Nair P, Aker N, Sampson EL, Moore K, Manthorpe J, et al. Rapid development of a COVID-19 care planning decision-aid for family carers of people living with dementia. Health Expectations: An International Journal of Public Participation in Health Care and Health Policy. 2022;**25**(4):1954-1966. DOI: 10.1111/hex.13552

[6] Emerson E, Fortune N, Llewellyn G, Stancliffe R. Loneliness, social support, social isolation and wellbeing among working age adults with and without disability: Cross-sectional study. Disability and Health Journal. 2021;**14**(1):100965-100965. DOI: 10.1016/j.dhjo.2020.100965

[7] Miskelly F. Gerontechnology: Growing old in a technological society edited by Gari Lesnoff-Caravaglia. Age and Ageing. 2007;**36**(6):706-706. DOI: 10.1093/ageing/afm125

[8] Sen K, Prybutok G, Prybutok V. The use of digital technology for social wellbeing reduces social isolation in older adults: A systematic review. SSM—Population Health. 2022;**17**:101020-101020. DOI: 10.1016/j.ssmph.2021.101020

[9] Kotwal AA, Fuller SM, Myers JJ, Hill D, Tha SH, Smith AK, et al. A peer intervention reduces loneliness and improves social well-being in low-income older adults: A mixed-methods study. Journal of the American Geriatrics Society (JAGS). 2021;**69**(12):3365-3376. DOI: 10.1111/jgs.17450

[10] Hung L, Liu C, Woldum E, Au-Yeung A, Berndt A, Wallsworth C, et al. The benefits of and barriers to using a social robot PARO in care settings: A scoping review. BMC Geriatrics. 2019;**19**(1):232-210. DOI: 10.1186/s12877-019-1244-6

[11] Yoshida N, Yonemura S, Emoto M, Kawai K, Numaguchi N, Nakazato H, et al. Production of character animation in a home robot: A case study of LOVOT. International Journal of Social Robotics. 2022;**14**(1):39-54. DOI: 10.1007/s12369-021-00746-0

[12] Olson KE, O'Brien MA, Rogers WA, Charness N. Diffusion of technology: Frequency of use for younger and older adults. Ageing International. 2011;**36**(1):123-145

[13] Davis FD, Bagozzi RP, Warshaw PR. User acceptance of computer technology:

A comparison of two theoretical models. Management Science. 1989;**35**(8):982-1003. DOI: 10.1287/mnsc.35.8.982

[14] Venkatesh V, Morris MG, Davis GB, Davis FD. User acceptance of information technology: Toward a unified view. MIS Quarterly: Management Information Systems. 2003;**27**(3):425-478

[15] Sumak B, Hericko M, Pusnik M. A meta-analysis of e-learning technology acceptance: The role of user types and e-learning technology types. Computers in Human Behavior. 2011;**27**(6):2067-2077. DOI: 10.1016/j.chb.2011.08.005

[16] Nelson EA, Dannefer D. Aged heterogeneity: Fact or fiction? The fate of diversity in gerontological research. Gerontologist. 1992;**32**(1):17-23

[17] Langdon P, Johnson D, Huppert F, Clarkson PJ. A framework for collecting inclusive design data for the UK population. Applied Ergonomics. 2015;**46**:318-324. DOI: 10.1016/j.apergo.2013.03.011

[18] Chen K, Chan AHS. Gerontechnology acceptance by elderly Hong Kong Chinese: A senior technology acceptance model (STAM). Ergonomics. 2014;**57**(5):635-652. DOI: 10.1080/00140139.2014.895855

[19] Farage MA, Miller KW, Ajayi F, Hutchins D. Design principles to accommodate older adults. Global Journal of Health Science. 2012;**4**(2):2-25. DOI: 10.5539/gjhs.v4n2p2

[20] Vance DE. Gerontechnology: Growing old in a technological society: By Gari Lesnoff-Caravaglia, PhD. Springfield, IL: Charles C Thomas Publisher, LTD, 2007. 314 pages. Illus., hardback, $69.95. Physical and Occupational Therapy in Geriatrics. 2008;**26**(4):69-70. DOI: 10.1080/02703180801963949

[21] Chen K, Lou VWQ. Measuring senior technology acceptance: Development of a brief, 14-item scale. Innovation in Aging. 2020;**4**(3):igaa 016-igaa016. DOI: 10.1093/geroni/igaa016

[22] Moyle W, Arnautovska U, Ownsworth T, Jones C. Potential of telepresence robots to enhance social connectedness in older adults with dementia: An integrative review of feasibility. International Psychogeriatrics. 2017;**29**(12):1951-1964. DOI: 10.1017/S1041610217001776

[23] Bernabei V, De Ronchi D, La Ferla T, Moretti F, Tonelli L, Ferrari B, et al. Animal-assisted interventions for elderly patients affected by dementia or psychiatric disorders: A review. Journal of Psychiatric Research. 2013;**47**(6):762-773. DOI: 10.1016/j.jpsychires.2012.12.014

[24] Moyle W, Cooke M, Beattie E, Jones C, Klein B, Cook G, et al. Exploring the effect of companion robots on emotional expression in older adults with dementia: A pilot randomized controlled trial. Journal of Gerontological Nursing. 2013;**39**(5):46-53. DOI: 10.3928/00989134-20130313-03

[25] Kyoko O, Hiroko S, Tetsuya T, Yuko Y, Rozzano L, Yueren Z, et al. Characteristics of a transactive phenomenon in relationships among older adults with dementia, nurses as intermediaries, and communication robot. Intelligent Control and Automation. 2017;**8**(2):111-125. DOI: 10.4236/ica.2017.82009

[26] Ke C, Lou VW-Q, Tan KC-K, Wai MY, Chan LL. Changes in technology acceptance among older people with dementia: The role of social robot engagement. International Journal

of Medical Informatics (Shannon, Ireland). 2020;**141**:104241-104241. DOI: 10.1016/j.ijmedinf.2020.104241

[27] Lee S. Variability in psychosocial correlates of health among the elderly: A comparison of the EU and US. European Journal of Public Health. 2016;**26**(Suppl. 1). DOI: 10.1093/eurpub/ckw164.057

[28] Cacioppo JT, Cacioppo S. Older adults reporting social isolation or loneliness show poorer cognitive function 4 years later. Evidence-Based Nursing. 2014;**17**(2):59-60. DOI: 10.1136/eb-2013-101379

[29] Edgar A, Bowling A. Measuring health; a review of quality of life measurement scales. In: Medicine, Health Care and Philosophy. 2nd ed. Vol. 1(2). 1998. pp. 181-182. DOI: 10.1023/A:1009999222296

[30] Russell DW. UCLA loneliness scale (version 3): Reliability, validity, and factor structure. Journal of Personality Assessment. 1996;**66**(1):20-40. DOI: 10.1207/s15327752jpa6601_2

[31] Lyubomirsky S, Lepper HS. A measure of subjective happiness: Preliminary reliability and construct validation. Social Indicators Research. 1999;**46**(2):137-155. DOI: 10.1023/A:1006824100041

[32] Triandis HC, Gelfand MJ. Converging measurement of horizontal and vertical individualism and collectivism. Journal of Personality and Social Psychology. 1998;**74**(1):118-128. DOI: 10.1037/0022-3514.74.1.118

[33] Lewis JR. The system usability scale: Past, present, and future. International Journal of Human–Computer Interaction. 2018;**34**(7):577-590. DOI: 10.1080/10447318.2018.1455307

Chapter 3

Exploring Key Challenges in Child-Robot Interaction Using *Haru4Kids*: Engagement, Language Understanding, and Privacy

Gonzalo A. García, Leigh M. Levinson, Guillermo Pérez, Manuel Castro, José Gabriel Amores, Gloria Álvarez, Randy Gomez and Selma Šabanović

Abstract

This chapter examines three critical challenges in long-term child-robot interaction in the home, once the novelty effect has faded away: engagement, language understanding, and privacy concerns. The study used the *Haru4Kids* (H4K) platform, a child-oriented family robot simulator that features a rotating iPad-based interface offering interactive activities. This platform facilitated sustainable child-robot interactions in family environments, as it offered an interactive platform while also allowing us to assess user engagement and behavior throughout each interaction. Over two weeks, seven families in Southern Spain cohabitated with H4K. The study provides comprehensive insights into user engagement by integrating user logs, annotated images, and verbal interaction analysis. The image-based engagement was assessed via an innovative Engagement Level Metric, which we used to estimate which activities the children found most engaging. A natural language processing analysis revealed that mixed-initiative dialogs enhanced user agency over time, shifting interactions from system-driven to user-driven. Privacy concerns varied between children and parents, with children showing more hesitancy toward third-party data sharing. This chapter offers valuable design recommendations for future child-robot interaction platforms, emphasizing personalization, transparent data practices, and diverse activity offerings. For researchers and developers, it underscores the importance of addressing dynamic and multi-modal engagement and privacy concerns in realistic, unsupervised settings.

Keywords: child-robot interaction, engagement estimation, natural language processing, dialog management, privacy concerns, study *in the wild*, novelty effect

1. Introduction

Child-robot interaction (CRI) is a field within human-robot interaction that is growing in popularity and interest. Robots are increasingly being developed for children in influence areas like education [1, 2] and therapy [3]. However, much CRI research is limited to controlled settings due to the absence of robust, unsupervised platforms. Unlike controlled lab environments, "in-the-wild" environments, like a family home, allow children to interact with social robots voluntarily and in more dynamic ways [4]. This chapter introduces the *Haru4Kids* (H4K) platform, designed intentionally as a test-bed to evaluate children's acceptance of robot *Haru* [5], particularly focusing on how children's engagement with H4K changes across activities and over time, and on gauging families' concerns about privacy.

Our work builds on previous studies involving Haru in child-robot interaction and aligns with UNICEF's *Policy Guidance*,[1] which addresses children's fundamental rights with AI and robots, including their right to privacy and fairness [6]. Designing a robotic system in line with a child's rights requires not only careful consideration of the contexts in which the robot is to be used but also the guiding framework and scope of the interactions. It also demands careful consideration of the design features that will sustain long-term and satisfying CRI.

In order to achieve these goals, H4K features an avatar of the robot *Haru* [5, 7] displayed on an iPad held by a rotating stand that orients the iPad to face the child as they move. The design principle guiding Haru emphasizes a balance among human expectations, appearance, and functionality [5, 8]. This design principle takes a more holistic approach to CRI, favoring a variety of research methods and study designs to learn about it.

In this chapter, we integrate the findings from three studies: (1) takeaways from children's engagement estimation by visual clues [9]; (2) different dialog strategies (user-driven, system-driven, and mixed-initiative) and evaluation of common built-in language intents of the system [10]; and (3) the pre- and post-perceptions of parents and children of information sharing and privacy concerns with robots in the home [11]. This work reflects on the fundamental preconditions for long-term child-robot interactions in dynamic environments like the home, including robot vision-based measurement of a child's engagement with the system, improving child-robot dialogs, and aligning robotic use with familial comfort with sharing information. All of these components represented in our work are understudied in child-robot interaction (CRI), although they add great value as foundations for more sustainable integration of robots in children's spaces.

2. Background

2.1 Long-term child-robot interaction

Cohabitating robots have been introduced to the home for a variety of purposes, such as entertainment, learning, healthcare, and companionship [12]. In the pursuit of measuring a *successful* and acceptable integration into children's spaces, the field looks at how robots are used over longer periods. Defining "long-term" depends on interaction frequency, quality, capability, diversity, and scope [13, 14].

The kinds of benefits that sustain children's interaction over time can include hedonic enjoyment or utilitarian benefits, like learning [15]. Additionally, the perception of a physical robot's social presence and ability to exhibit expressive motion

[1] https://www.unicef.org/globalinsight/media/2356/file.

enhances user engagement compared to virtual agents [16, 17]. The personalized nature of robots is also a desirable feature and can enhance long-term engagement [18]. Furthermore, children are interested in robots that are expressive across verbal and non-verbal channels of communication, customizable, relatable, and approachable [19].

While establishing relationships enhances sustained interactions [20], doing so in the long term is challenging. Long-term studies show that many robotic platforms cannot overcome the *novelty effect*—the amplified but temporary excitement and use of a new device [12, 13]. As such, it is increasingly important for the field to define and measure different channels of successful interaction. In the following sections, we will expand on methods for measuring non-verbal and verbal forms of engagement with a robotic platform, and also the privacy concerns that may hamper even the short-term use and acceptance of robots.

2.2 Defining and measuring engagement

As already discussed, engaging and developmentally appropriate activities can support sustained interactions tailored to individual needs [14]. However, measuring the engagement of a user during an interaction is not a trivial task. Engagement, as defined by O'Brien [21], is the cognitive, affective, and behavioral investment in digital interaction, measurable through self-reports like the User Engagement Scale [22] and automatic estimation methods. Engagement with neurodivergent populations further suggests diversifying "engagement" across contexts, as it can support more personalized interactions over time [23]. The dynamic nature of engagement is also found in the home, for instance, with extended studies of home-based robots as reading companions demonstrating children's evolving engagement over several months [24, 25].

Currently, the most reliable engagement measurement methods rely on physiological signals [26], but these are intrusive and cumbersome for children. Other measurements of engagement are visual-based or multi-modal, which are unobtrusive and cost-effective [27], such as the recent one based on thermal imaging [28]. The existing automatic engagement estimation methods can be divided into those based on feature extraction and the ones that perform automatic prediction [29].

In our work, we look at visually-based engagement, including automatic face-angle tracking and hand-annotated engagement levels, which provide high resolution with minimal bias [30, 31]. Additionally, we explored engagement in CRI through dialog analysis, which is more objective, but less detailed.

Our approach uses a human-in-the-loop model, which integrates the feedback of human annotators for long-term engagement measurement in CRI. In this way, we take advantage of the natural human skill of *face reading* [32].

2.3 NLP as a key factor for engaging interactions

Dialog, enabled by Natural Language Processing (NLP), is a key way of achieving verbal engagement with young users. NLP has advanced significantly, making it possible for robotic systems to carry out an increasingly natural dialog with users. However, children's smaller vocal tracts and evolving language use create distinct difficulties in speech recognition [33, 34]. The lack of large, children-specific datasets further complicates speech recognition development, which is generally dominated by large corporations [35]. Children's imaginative language, ungrammatical phrases, and unique interests pose additional challenges [36]. Our work examines these issues in children's homes, an under-researched setting in CRI studies [37].

Unlike embodied robots, conversational agents have been more extensively studied in the home. Research on Alexa (Amazon), for example, explores breakdowns and repair strategies in communication [38] and analyzes language processing for joint reading [39]. These findings underscore the need for adaptive interaction models in child-robot dialog systems. As such, we pursue not only visual engagement analysis but also analyze the flow of dialogs with children and ways to improve verbal engagement with at-home robots.

2.4 Family privacy concerns

Though it is an understudied topic—surely due to its intrinsic difficulty to grasp and assess—ensuring privacy with at-home and child-centered robots is desired by family users [18]. Children's right to privacy builds from the general definition of privacy as the right to control what a user shares with another agent and how accessible what was shared is to others [40]. Dimensions of privacy also consider the control over being physically or psychologically alone [41]. As the kind of information that is shared varies in nature, it is expected that a user's willingness to share information or their privacy concerns will change based on the change of what is shared, whom it is being shared with, and how it is shared. This contextual nature of comfort is named by the framework of *contextual integrity* [42]. Within the context of social robots, we focus on informational, physical, social, and psychological dimensions of privacy given the heightened emotional and functional cues leveraged during interaction [43].

As a user's engagement shifts dynamically, their fluid relationship with robots across different contexts leads them to implicitly and explicitly navigate boundaries that they and their families establish. This boundary management has been described by the *communication privacy management theory*, which recognizes how users make others "co-owners" of their information depending on when they grant them access to information under their control [44]. In the case of families, pre-existing power structures, systematic hierarchies, and emotional relationships add a layer of complexity to privacy concerns [45].

These concerns can also vary widely across user demographics. Individual and generational differences affect how families establish, maintain, and amend these boundaries as they have done with other technologies in the past [46]. The privacy paradox with robots also recognizes how perceptions and expressed discomforts differ from actual user behavior [47]. All this context-driven variability in the management of privacy inspired our study of how parents and children conceptualize their privacy with a cohabitating social robot. Furthermore, studying familial privacy concerns will help researchers navigate the best contexts for the most comfortable engagement with these systems in the home.

3. Participating families and methods

3.1 Participant families

Seven families from Southern Spain participated in our study, recruited through convenience sampling. The study was approved by Indiana University's Institutional

Review Board #14363, with caregivers providing written consent and children (aged seven and older) giving written assent. In total, H4K engaged with 14 children (nine male) between ages 6 and 13 (mean age = 9.6). This sample size aligns with prior exploratory CRI studies, which often involve small samples, as noted in related works (e.g., [48–50]). While one family had an only child, the other families included at least two children. Accounting for periods when children were away, H4K cohabitated with each family for an average of 20 days ($\sigma = 5.9$, max = 27), resulting in a net period of two weeks per family. Researchers visited each home to install and set up H4K, explain privacy configurations to caregivers, and introduce the platform to family members. Children were then free to explore H4K independently.

3.2 *Haru4Kids* platform

H4K consists of a Haru avatar displayed on an iPad supported by a rotating stand (**Figure 1a**), using Apple libraries across seven modules: User Interface, Conversation Manager, Vision Manager, Central Controller, Settings Manager, Logging, and Stand Manager (**Figure 1b**). The Conversation Manager leverages cloud-based services for certain functions, while the Central Controller coordinates app behaviors and cloud access as needed. H4K's rotating stand augments its embodied presence, differentiating it from other virtual or voice-based systems.

H4K offers several activities designed for engagement and cognitive stimulation: *Storytelling*, *Gusano Loco* (a humorous, word-based storytelling activity), *Detectives* (word matching), *Would You Rather ... ?* (choosing between silly options), and *Jokes*. These activities, created with neuropsychologists and speech therapists, were designed to stimulate different cognitive functions in children (for further details, see [9]).

3.3 Data collection

We collected data before, during, and after cohabitation. Images, audio, and logs were collected throughout the interaction. For a more holistic understanding of information sharing and general feedback on using the platform, we interviewed family

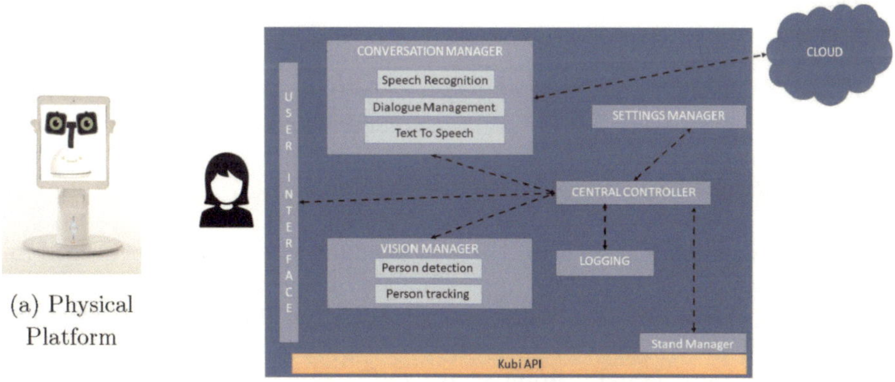

(a) Physical Platform

(b) Technical Description

Figure 1.
Haru4Kids: *(a) iPad supported by rotating-tabletop stand, and (b) high-level technical description of the system.*

members before and after they interacted with H4K. Below, we describe the data collected in greater depth. All data was stored on the iPad locally and collected when the robots were retrieved from families' homes.

3.3.1 Experiment logistics

The families were recruited among acquaintances of one of the co-authors. The exact same protocol was followed with all the families: first, we made an appointment for deployment by e-mail. All deployments were carried out within the same week to keep the experimental conditions as similar as possible among the households. During the deployment, the same two researchers followed an identical process in all the homes: provide general explanations; provide explanations of the research consent, assent, and of the *General Data Protection Regulation* (GDPR) documents, first to parents, then to children: allow participants to read and sign the consent, assent, and GDPR documents; collect demographic data; complete pre-interview (parents, then children, separately); place the platform where the family indicates; set up the system; do an initial check; give brief instructions to family about how to use the system. Participants were given the contact information of the two researchers in charge of the experiment and encouraged to reach out if there were any issues.

After two net weeks, the researchers picked up the H4Ks and conducted a post-interview with the family members. The data was uploaded to the researchers' local server, which is highly protected and is only accessible to the researchers. The data were then permanently and irrecoverably deleted from the iPads.

3.3.2 User logs

All application events were logged both on the iPad and—anonymously—in the cloud (AWS Cloudwatch). Log entries included timestamps, event types (e.g., actions by H4K or user requests), user head angles (from Apple's Vision framework), and rotating stand movements, as these can correlate with engagement levels [51]. These logs provide a continuous record of user-robot interaction, supporting subsequent analysis of dialog quality [52].

3.3.3 Image capture

To gauge user engagement, we captured images of users' faces at one frame per second during interactions from six of the seven families who consented to image collection. Images were displayed in real time on the iPad for transparency.

3.3.4 Audio and conversation manager

H4K's Conversation Manager, in charge of handling the components needed for voice-based interactions, integrates Apple libraries for speech recognition and its synthesis (Text to Speech). The Dialog Management component is a custom-built container hosted on AWS that allows complex, mixed-initiative dialogs beyond standard chatbot interactions and features error-handling capabilities like *Repeat*, *Help*, *Background*, and *Sleep*, providing a robust interaction experience even in error-prone, unsupervised home environments.

Conversation Manager is also in charge of the audio signal recording, which was limited to interactions where children responded verbally, with data stored locally and never uploaded to the cloud. Families retained full control over data deletion at any stage of the experiment (even after it was finished), respecting privacy guidelines.

3.3.5 Pre- and post-interview data

To better understand caregiver and child experiences and concerns with a cohabitating robot, we also utilized pre- and post-interviews. In a 30-minute pre-interview, researchers asked children about their first impressions and expectations for interactions with the robot. Notably, we asked multiple-item questions about information sharing with a social robot (e.g., *"What kind of information would you feel comfortable showing or telling with a robot?"*) and about sharing information with different third parties through the robot (e.g., *"What information would you feel comfortable having the robot share with your teacher/friend/sibling/parent/robot creator?"*). Parents were asked about their own comfort with sharing information with the robot and comfort with their child interacting and sharing information with the robot and third parties (total of 7 closed, information-sharing questions with 12 items each).

At the end of the trial period, a 30–45 minute post-interview was conducted with all family members, who were asked the same multi-item questions for a pre-post analysis. Kids were also asked about what they enjoyed, did not enjoy, or would like Haru to do in the future. Parents were also asked about their impressions of cohabitation and feedback for the roboticists. Furthermore, we asked each participating child to draw Haru, so that we could gauge what they perceived as the most salient features of the robot.

For more specific details about the methods here presented, please refer to our publication on this study [11].

3.4 Data analysis

3.4.1 Interview analysis

Comfort levels and other quantitative measures were analyzed statistically using t-tests and a p-value of 0.05. Qualitative answers are described generally and reported without thorough thematic analysis. These interviews were mainly used for internal feedback and to guide the engagement analyses here described.

3.4.2 Measures for engagement estimation based on usage and dialog

From the log files, we want to call attention to the extraction of the usage profile of each family and user reactions to the robot. Specifically, for each family and each day of the experiment, we calculated the number of sessions per day and their durations based on their starting and ending timestamps. Additionally, we determined the number of times each activity was executed. We also analyzed whether it was actively requested by the user (user-driven) or proposed by the robot (system-driven) and whether it was completed or aborted by the user. From the log entries related to dialogs, we analyzed the users' reactions, or lack thereof, to the robot's participation cues to the user, so that we obtained an additional estimation of their level of attention and engagement.

3.4.3 Engagement profile estimation from users' pictures analysis

An *Engagement Level Metric* (ELM) was developed based on hand-annotated user photos categorized into four levels of engagement. Annotators, trained in a seminar and equipped with a guide, assessed a total of around 20 K collected images using the ELM, which follows descriptions from prior studies [53, 54]. Annotators were provided only with facial images, ensuring unbiased assessments. Our annotation support application monitored annotator attentiveness by inserting test images and tracking response times.

3.4.4 Additional measures for engagement estimation

- *User head angle (UHA)*: Correlation studies were conducted between UHA components and the ELM, as previous literature has associated engagement with angles within $\pm 12^{\circ}$, indicating user focus on the screen [51].
- *Stand cumulative angle (SCA)*: This metric measures the stand's movement, assuming a higher SCA indicates decreased engagement.

3.4.5 Assessing annotation quality and reliability analysis

Annotation reliability was assessed using inter-rater coefficients such as Krippendorff's α, Cohen's κ, and Cronbach's α [55–57], as well as general agreement measures like Pearson's correlation and Root Mean Square Deviation (RMSD), which was normalized by $(y_{max} - y_{min})$, 3 in our case.

4. Results

4.1 General use over time

Figure 2 shows the cumulative usage time across families over two weeks. Individual families' usage is shown in the inset of **Figure 2**. Overall usage was low, yielding a total of 18.37 hours, and quite variable both in usage length, with one family achieving 15 days of net use and another just three; and in total usage time, with a maximum of 4.92 hours and a minimum of 53 minutes. After day #10, only one family continued interacting with the robot, showing how once the novelty effect fades away, the usage drops steeply.

Figure 3a illustrates usage by activity, showing how children preferred *Jokes*, *Storytelling*, and *Gusano Loco*, which is aligned with self-reported preferences [11]. The interviews revealed results in line with this feedback. Of all the activities, children expressed the most excitement over jokes and the least excitement for Haru's storytelling.

In general, most interactions were user-driven (red bars), with children increasing activity requests from 33% on day one to 85% by the last day as shown in **Figure 3b**. This result shows the importance of the mixed-initiative approach, which allows the children to either passively wait for the robot's proposals, or to actively request a specific activity. Further details about this approach will be discussed in Section 4.4.2.

4.2 Visually measured engagement profile

Children's attentiveness was analyzed using verbal responses to H4K's prompts, with results shown in Section 4.4. Engagement levels were primarily assessed through

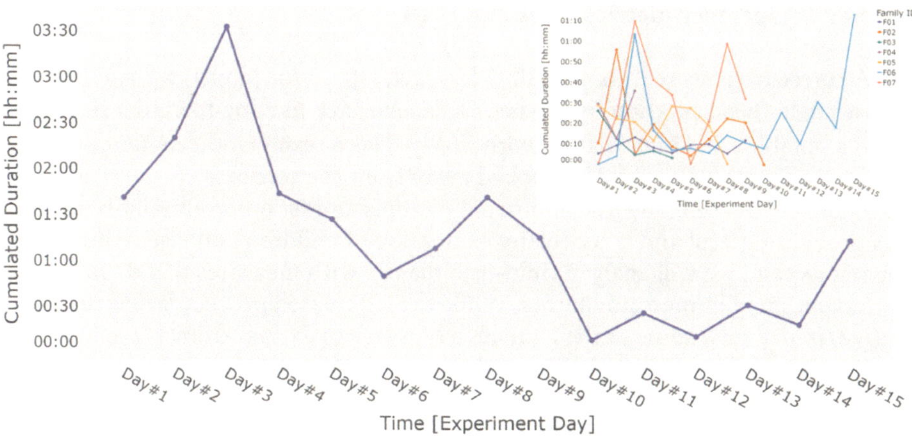

Figure 2.
Evolution of the average, global time usage by day of experiment. Inset: individual family behavior.

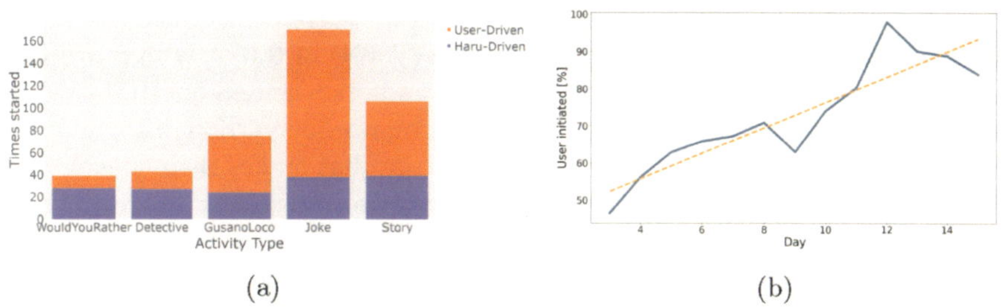

Figure 3.
Comparison of activity usage and user-initiated activities over time. (a) Comparison of activities usage. (b) User-initiated activities percentage over time.

hand-labeled interaction photos, though these were less objective than dialog-based metrics.

After annotating, the inter-annotator agreement was assessed, resulting in the exclusion of two annotators with low agreement scores. Inter-rater agreement level was mixed: Krippendorff's α indicated low reliability, while Cohen's κ suggested fair-to-moderate agreement, and Cronbach's α indicated good agreement. Those measures are known to underestimate agreement when one label dominates [58]. On the other hand, Pearson's r scores (0.43–0.55) were strong by psychological research standards [59]. Collectively, these values support the ELM's validity as an engagement estimate, given also the high RMSD-based metric (71–74%).

Based on those fair-to-moderate agreement values among the remaining 10 annotators, the data provide reasonable confidence in the labels used for engagement analysis. To further ensure the validity of the manual labeling, we selected from the labels for each set only those in which at least three annotators agreed in its ELM, which was 79% of the pictures. Afterward, we considered the ELM of each picture the most-voted label, $\overline{ELM}_3$.

Those metrics show that activities like *Detective*, *Would You Rather*, and *Gusano Loco* had slightly higher engagement. Engagement declined over time, but the differences among activities remained statistically significant, with *Detective* ranking highest and *Story* lowest, which is in line with the results obtained from user logs and the interviews.

4.3 Other engagement metrics: SCA and UHA

SCA correlations were minimal (0.07 for yaw, −0.03 for pitch), while UHA absolute yaw angle (head rotation by the neck axis, eye gaze moving toward a shoulder) showed a notable −0.43 correlation with $\overline{ELM}_3$. These results indicate that perceived engagement decreased as children looked away from the screen.

Overall, we describe how the annotated pictures amidst other visually-based indicators provide a useful and trustworthy estimation of children's engagement. In the following section, we will analyze child-user dialogs with measures that do not involve the intervention of human annotators. While the automated process makes them more objective, it may miss some of the nuances a human can extract from a face image. We recognize these differences and present the general takeaways from the study.

4.4 Child-robot dialogs and verbal engagement

4.4.1 Children's language understanding in the wild

The Automatic Speech Recognition (ASR) outputs were transcribed and analyzed by humans, yielding an overall Word Error Rate (WER) of 0.077, with multiple-choice questions performing best (WER = 0.07) and open-ended questions worse (WER = 0.11). Hence, ASR showed robust performance, even though some of the participating children have a strong Southern accent that makes their speech more difficult to interpret, as it differs from the *standard* pronunciation. In any case, the accuracy of the ASR could be further improved by optimizing for common names or fine-tuning the models.

Table 1 details intent recognition metrics, with high performance across all categories (F $(\beta = 0.5) \geq 0.97$). However, low occurrence counts for some intents (e.g., *Repeat*) may affect generalizability. The results exclude recognition errors, silences, and out-of-scope (OoS) responses, but these collectively accounted for 24% of responses in yes/no and multiple-choice questions.

Handling OoS inputs effectively is vital for user experience [60], with Dialogflow's fallback intent yielding a performance of $Precision = 0.90$ and $Recall = 0.98$.

Error rates declined slightly over time, with "silence" rates decreasing markedly toward the end of the trial, suggesting children became more familiar with how to respond to the robot's prompts.

4.4.2 Children-robot dialog management in the wild

Standard dialog metrics, such as average user turns (5.25) and session duration (1.92 minutes), varied between families, with one family averaging almost five

	yes/no	multiple	sleep	repeat	activity
Precision	0.992	0.985	1.000	1.000	0.993
Recall	0.985	0.958	0.886	1.000	0.987
F $(\beta = 1)$	0.989	0.971	0.939	1.00	0.990
F $(\beta = 0.5)$	0.991	0.980	0.975	1.00	0.992

Table 1.
Intent recognition performance.

minutes per session. As already shown in **Figure 3b**, user-requested activities increased over time, indicating growing familiarity with the system.

Domain-independent intents (e.g., *Repeat* and *Help*) were infrequently used, except for *Background*, which was frequently triggered (1.31 times per session). Most sessions ended automatically when users disengaged, with fewer instances of using the *Sleep* intent manually. The NLU layer performed well in general, with pre-defined intents achieving F-scores of 0.98 or higher. While recognition issues affected 10% of inputs, they did not decrease over time, suggesting this may be a stable limit for our current setup.

Within this work, our takeaways are two-fold: (1) there is a clear benefit of mixed-initiative dialog, allowing users to shift from predominantly system-driven interactions to increasingly user-driven interactions by the experiment's end; and (2) the observation that functionalities like *Help* and *Repeat* require explicit user training, while a fallback *Background* strategy is critical in maintaining flow in conversational applications.

Further, H4K offers valuable dialog statistics on turn-taking, error handling, and dialog closure (see details in [10]) that could inform future CRI designs.

4.5 Dynamic comfort in sharing information in the home

To complement the above metrics of visual and verbal engagement with the robot, we report the evolution of the family members' general comfort of having Haru in the home. See **Table 2** which describes children's change in comfort in sharing

Info Category	General	Third Parties				
		Teacher	Friends	Siblings/ Cousins	Parents	Robot Creators
Third Party		74%	80%	83%	86%	40%
School Grades	57% ↓3	100%	50%	50%	50%	21% ↓3
Hobbies	100%	79% ↓1	100%	93% ↓1	86% ↓1	86%
Conversations with Others	14% ↓2	7% ↑1	14% ↓1↑1	14% ↓3↑1	29% ↓3↑1	14% ↓1
Name	100%	100%	100%	100%	100%	42% ↓3
Birthday	86% ↓2	93% ↓1	100%	100%	100%	21% ↓5
Pets	100%	93%	100%	100%	100%	71% ↓1
Family Info	43% ↓1	35% ↓1	50% ↓1	100%	93% ↓1	14% ↓2↑1
Friend's Info	64% ↓3	93% ↓1	100%	50% ↓3	79% ↓2	42% ↓4
Location	43% ↓3	36% ↓1	76% ↓1	100%	100%	21%
Voice Recognition	100%	100%	100%	100%	100%	71% ↓4
Face Recognition	93%	100%	100%	100%	100%	42% ↓3
Images of family	57% ↓2	42% ↓3	57% ↓3↑1	83% ↓1	93% ↓1	21% ↓2

↓ = *number of children who changed to discomfort in the post-interview.*
↑ = *number of children who changed to comfort in the post-interview.*
Blue = 100% Comfort, Green = ≥50% comfortable, Red = <50% comfortable.

Table 2.
Child comfort with sharing information originally reported by Levinson et al., 2022 [11].

information. All children, regardless of previous exposure to a prototype, had high initial expectations for the robot's capabilities. As such, children reported disappointment when H4K struggled to understand their input, which was related to our NLP findings and impacted interaction frequency. There was also an overall decrease in comfort with sharing information, in line with this reported disappointment in the robot's capabilities.

Regarding privacy, only 14% of children felt comfortable sharing conversations with others with the robot, and 42% felt comfortable sharing location or family information with the robot. The most comfortable information shared included names, pets, and voices.

Post-interview results showed that 98% of adults were comfortable with children sharing data directly with them but were less at ease with sharing data with teachers (69%) and robot creators (67%), underscoring the contextual comfort with whom information is shared. In comparing children's pre- and post-interview results, a paired t-test determined that there was a significant difference before and after the cohabitation period ($t = 5.303$, $p < 0.001$). In comparison, a paired t-test of the adult's pre- and post-interview responses about information sharing was not significant ($t = 1.387$, $p = 0.22$). Therefore, children were more likely than parents to change their answers post-interaction than their caregivers. While this does not directly reflect on the privacy paradox, which identifies how user behavior differs from user preferences, it elucidates that younger children have more dynamic boundaries, justifying a greater need to be transparent with information sharing so they can make informed decisions on their robot use.

5. Discussion

To our knowledge, our work is the first one to provide such a multi-faceted study in CRI in the home environment. In the following discussion, we will situate our work within the research on CRI and the deployment of engaging systems.

5.1 Children's dialogs with robots

As Ljunglöf et al. [61] noted, speech recognition has historically been dialog systems' weak point, but our results align with recent improvements in ASR technology [62]. Although not yet at human-level accuracy, these results show notable progress, particularly for non-English languages like Spanish in real-world contexts.

Our reliable performance of speech recognition in this challenging scenario, similar to the accuracy rates reported in the study of Xu and Warschauer [39], indicates progress but leaves more to be desired. There are challenges unique to voice interfaces with children, such as children's reliance on and use of non-verbal gestures in communication [39], which motivates an enhanced focus on dialog management for child-robot interactions with multi-modal capabilities.

As for language understanding, although our NLU layer performed well in general, new NLUs based on current Large Language Models (LLMs) could overcome some limitations of more "classic" approaches like ours, as LLMs offer a deeper apprehension of user inputs. However, LLMs arise some issues such as privacy concerns and the possibility of *hallucinations* and non-proper answers.

Mixed-initiative dialogs allowed children to initially follow system-driven interactions and later take control to request specific activities, as shown in **Figure 3b**. This is a novelty with respect to existing systems, which are either fully passive—just answering the questions addressed to them by the user (e.g., *Alexa*)—or fully autonomous—guiding the user through the (maybe pre-established) dialog (e.g., a voice menu over a phone call). Mixed-initiative dialogs allow users to become more familiar with the system by the scaffolding of the guided dialog until they feel confident enough to lead the dialog.

5.2 A multi-modal engagement study

Our overall work is comprehensive in that it can take into account engagement measures across user logs and annotated images, dialog responses, and self-report interviews. Visual engagement estimation and interviews were performed on an individual's interaction data, whereas estimations by dialog and usage were rather measured family-wide. Our results found that generally, children and parents, while initially comfortable with sharing images during interactions, became slightly less comfortable after the interaction. However, they were more comfortable with voice recognition and the recording of audio data. While images may offer a rich and well-established source of information about the interaction, these results inspire future work on triangulating engagement across these modalities to best align data collection with familial comfort.

It is worth noting that the ELM ranking of activities $(\text{Detective} \gtrsim \text{WYR} \approx \text{Jokes} \gtrsim \textit{Gusano Loco} \gtrsim \text{Story})$ collected from images did not correspond directly to usage frequency from the user logs or self-reported enjoyment $(\text{Jokes} > > \text{Story} > > \textit{Gusano Loco} > \text{Detective} \approx \text{WYR})$. In this way, we identify an inequality between the activities most enjoyed as reported by children and their families, most frequently used, and those in which children appeared more engaged. Therefore, ELM could guide design for highly engaging activities, while usage profiles could identify the characteristics of popular, frequently used content. This approach offers a dual perspective for activity design in future CRI platforms.

This multi-modal approach also validated diverse kinds of engagement in CRI. The higher engagement levels of *Detective* could inform the design of future activities that emphasize problem-solving and interaction diversity. In fact, studies have highlighted that game-based problem-solving activities foster intrinsic motivation and sustained engagement in children during human-computer interaction scenarios, aligning with our findings [63]. Additionally, the decline in engagement after the novelty effect wanes indicates that personalization and adaptability—features where mixed-initiative dialogs play a role—are essential to sustaining interest. On the side of personalization, we highlight how our Engagement Level Metric (ELM) provided a robust framework for estimating engagement based on image annotations. While tools like ELAN [64] offer annotation capabilities, our custom tool was specifically designed for this study's requirements, including features for annotator performance monitoring in real time and future usability. In this way, the engagement metric is more personalized to the participating children.

Our results are also in line with other work that emphasizes the role of mixed-initiative for engaging human-robot interactions [65], particularly in the case of maintaining engagement over longer periods of time [66].

Furthermore, we realized that children are not always more engaged by the activities they requested the most, as at some ages, they like serial repetitions of known content [67]. Thus, we recognize that both the nature of the interaction and the content are important contributing factors toward engaging CRI.

5.3 Privacy and information sharing

Beyond takeaways surrounding the topic of engagement, we briefly report some of the privacy-related findings that establish a foundation for trusting and successful child-robot relationships. Survey responses revealed an overall decrease in comfort in sharing information with a robot, with children exhibiting more caution than parents around third-party data sharing. The dynamic shift and decrease after cohabitation also make space for the re-navigation of privacy boundaries in the home, in line with the privacy communication management theory proposed before [44]. In this, we also find that children are more aware of the privacy risks with technology than parents or other adults realize [40].

In particular, we found that voice recognition was the most accepted information category among all dimensions, especially sharing with robot creators. With this in mind, it becomes critical to shift the CRI research focus toward improving dialog-based NLP and using voice-based metrics to assess the quality of child-robot interactions.

5.4 Limitations and future work

Our findings align with past research indicating challenges in sustaining children's attention over time, as the novelty effect wears off [68]. Activity decreased gradually. Limited content and the lack of embodiment might have influenced engagement, with our app-based platform lacking the physical presence that some children might prefer in social agents [11]. While embodied agents often offer benefits over screen-based ones [69, 70], we used an avatar to allow faster, more flexible deployment in multiple homes. This setup successfully ensured technical consistency throughout the experiment, reducing the risk of hardware issues and enabling easy configuration adjustments without compromising activity design. Though our platform has a rotating stand providing a degree of embodiment and movement, we recognize that this may have been a factor in decreased interaction. In the future, we hope to best address this with more robust platforms that can be both consistent and embodied.

This study was limited by sample size, language, and geographic scope. A planned follow-up will expand these areas, including longer cohabitation, broader demographics, and more diverse activities. Key enhancements include adapting activities to various developmental stages and cultural contexts and introducing collaborative storytelling. We plan to develop additional educational resources and explore cohabitation with other populations, such as hospitalized children, to leverage H4K's cognitive benefits for diverse users [71].

The two-week cohabitation demonstrated initial user interest that tapered during the second week, likely due to the onset of summer break. Future deployments should address expectations with transparency and age-appropriate content. In addition, the nearly 20 K pictures annotated in this study could be used as a dataset for training artificial intelligent models aimed at automatic engagement estimation, even in real time, closing the loop and making it possible to adapt the behavior of the robot to the attention level detected in the child user.

The intents *Help* and *Repeat* were underused, likely due to limited awareness of their existence. Therefore, we will need to introduce these features more explicitly when deploying the robots in order to improve future interaction.

As for privacy issues, future research and developments will have to take into account new legislation such as the European *AI Act*,[2] which places limitations on the collection of some types of information, as well as the use of large language models (e.g., *Meta*'s *Llama* 3.2[3]). One possible workaround for this could be to create more complex and useful logs, in which interpretations of the acts performed by the user—rather than the recording of the acts themselves—will be locally recorded in the robot and then used by the researchers, who will thus not have access to identifiable, private information (audio or video signals). Such an approach is already in practice in devices like the one used by the USA national nonprofit *LENA* (*Language ENvironment Analysis*): a small wearable device (a "talk pedometer") that records a log of events (e.g., the number of turns in a dialog) rather than the dialogs themselves and that it is used for speech analysis of toddlers and children.

6. Conclusions

The Haru4Kids (H4K) platform presented in this chapter represents a significant contribution to the field of child-robot interaction (CRI), providing a robust and versatile tool for studying engagement, language understanding, and privacy concerns in real-world, uncontrolled settings. By integrating multi-modal engagement metrics —ranging from visual data and dialog analysis to self-reported feedback—this research underscores the complexity of fostering meaningful and sustained interactions between children and social robots.

Our findings reveal critical insights: children exhibit enthusiasm for cohabitating robots, though their engagement fluctuates based on activity type, system familiarity, and the waning novelty effect. Privacy concerns emerged as another pivotal dimension, with children and parents navigating shifting comfort levels over time. In our opinion, these observations highlight the need for adaptive and transparent CRI systems that address both ethical and technical challenges.

The study's novelty lies in its comprehensive approach: combining user-driven and system-driven dialogs through mixed-initiative strategies, leveraging annotated image-based engagement metrics, and exploring privacy dynamics in unsupervised family environments. These contributions provide a foundation for future advancements in CRI, particularly in personalizing interactions and ensuring ethical data practices. In addition, our results may be useful for studies in developmental psychology and user engagement, supporting a growing field of research that combines objective and subjective engagement measures to foster meaningful, interactive child-robot experiences.

The broader implications of this research lie in its potential to guide the development of incoming child-oriented systems that balance entertainment, education, and ethical considerations [72].

Looking ahead, we propose enhancing activity diversity and platform embodiment while expanding demographic and cultural diversity in future trials. We also emphasize the integration of advanced natural language processing systems to deepen interaction quality. This research not only advances scientific understanding of CRI but

---

[2] https://www.europarl.europa.eu/topics/en/article/20230601STO93804/eu-ai-act-first-regulation-on-artificial-intelligence.

[3] https://www.llama.com/llama3_2/use-policy/.

also provides actionable insights for developing engaging, ethical, and inclusive child-robot interaction platforms for diverse real-world applications.

Acknowledgements

We would like to thank all the participating families, Sol Benitez Casaccia for helping with the translations throughout the process, and Sunzia Reza, Margaret Nigh, and Kaleb Lewellen for their help with content generation. The authors also thank Marta López, Ricardo Durán, Iván Delgado, José Andrés Millán, and Douwe Gelling for their technical essential contributions.

This publication is part of the project *PLEC2023-010251*, funded by MICIU/AEI/10.13039/501100011033.

This work was supported by *Honda Research Institute Japan* and by the Spanish Ministry of Science, Innovation and Universities under grants *PTQ2021-011726* and *PLEC2023-010251* funded by MICIU/AEI/10.13039/501100011033.

ChatGPT 4o (*OpenAI*) was used in the elaboration of this manuscript for style polishing, improving the tables, and summarization of some sections.

Author details

Gonzalo A. García[1,2*], Leigh M. Levinson[3], Guillermo Pérez[1], Manuel Castro[1], José Gabriel Amores[4], Gloria Álvarez[4], Randy Gomez[5] and Selma Šabanović[3]

1 4i Intelligent Insights, Seville, Spain

2 UNIR - Universidad Internacional de La Rioja, Logroño, Spain

3 Indiana University Bloomington, Bloomington, USA

4 Universidad de Sevilla, Seville, Spain

5 Honda Research Institute, Wako, Japan

*Address all correspondence to: g.garcia@4i.ai

References

[1] Toh LP, Causo A, Tzuo PW, Chen IM, Yeo SH. A review on the use of robots in education and young children. Journal of Educational Technology and Society. 2016;**19**(2):148-163

[2] Belpaeme T, Kennedy J, Ramachandran A, Scassellati B, Tanaka F. Social robots for education: A review. Science Robotics American Association for the Advancement of Science. 2018;**21**(3):eaat5954

[3] Cabibihan JJ, Javed H, Ang M, Aljunied SM. Why robots? A survey on the roles and benefits of social robots in the therapy of children with autism. International Journal of Social Robotics. 2013;**5**(4):593-618

[4] de Graaf M, Ben Allouch S, van Dijk J. Why Do they Refuse to Use my Robot? Reasons for Non-use Derived from a Long-Term Home Study. New York, NY: ACM Press; 2017. pp. 224-233

[5] Gomez R, Szapiro D, Galindo K, Nakamura K. Haru: Hardware design of an experimental tabletop robot assistant. International Conference on Human-Robot Interaction (HRI). 2018:233-240

[6] Charisi V, Imai T, Rinta T, Nakhayenze JM, Gomez R. Exploring the concept of fairness in everyday, imaginary and robot scenarios: A cross-cultural study with children in Japan and Uganda. Interaction Design and Children. 2021;**24**:532-536

[7] Sandry E, Gomez R, Nakamura K. Art, design and communication theory in creating the communicative social robot 'Haru'. Frontiers in Robotics and AI. 2021;**8**:577107

[8] Ackerman E. Haru: An experimental social robot from honda research. IEEE Spectrum. 2018. Available from: https://spectrum.ieee.org/haru-an-experimental-social-robot-from-honda-research

[9] García GA, Pérez G, Levinson L, Amores JG, Alvarez-Benito G, Castro-Malet M, et al. Living with Haru4Kids: Study on children's activity and engagement in a family-robot cohabitation scenario. In: 32nd IEEE Conference on Robot and Human Interactive Communication (RO-MAN). New York, USA; IEEE; 2023. pp. 1428-1435

[10] Pérez G, García GA, Castro M, Castaño M, De Quevedo MJLG, Durán R, et al. Child-robot conversation in the wild wild home: A language processing user study. In: 2023 32nd IEEE International Conference on Robot and Human Interactive Communication (RO-MAN). New York, USA; IEEE; 2023. pp. 1436-1442

[11] Levinson L, García GA, Pérez G, Alvarez-Benito G, Amores JG, Castaño-Ocaña M, et al. Living with Haru4Kids: Child and parent perceptions of a co-habitation robot for children. In: Social Robotics: 14th International Conference, ICSR 2022, Florence, Italy, December 13–16, 2022, Proceedings, Part II. Cham: Springer Nature Switzerland; 2023. pp. 54-63. DOI: 10.1007/978-3-031-24670-8_6

[12] Leite I, Martinho C, Paiva A. Social robots for long-term interaction: A survey. International Journal of Social Robotics. 2013;**5**(2):291-308

[13] Ahmad MI, Mubin O, Orlando J. Adaptive social robot for sustaining social engagement during long-term children–robot interaction. International Journal of Human Computer Interaction. 2017;**33**(12):943-962

[14] Cagiltay B, White N, Ibtasar R, Mutlu B, Michaelis J. Understanding factors that shape children's long-term engagement with an in-home learning companion robot. In: Proceedings of the 21st Annual ACM Interaction Design and Children Conference. New York, USA: ACM; 2022. pp. 362-373

[15] De Jong C, Peter J, Kühne R, Van Straten C, Barco A. Exploring children's beliefs for adoption or rejection of domestic social robots. In: 2021 30th IEEE International Conference on Robot & Human Interactive Communication (RO-MAN). New York, USA; IEEE; 2021. pp. 598-603

[16] Leite I, Martinho C, Pereira A, Paiva A. As time goes by: Long-term evaluation of social presence in robotic companions. In: RO-MAN 2009 - the 18th IEEE International Symposium on Robot and Human Interactive Communication. Toyama, Japan: IEEE; 2009. pp. 669-674

[17] Knight H, Simmons R. Expressive motion with x, y and theta: Laban effort features for mobile robots. In: 23rd IEEE International Symposium on Robot and Human Interactive Communication. New York, USA; IEEE; 2014. pp. 267-273

[18] Cagiltay B, Ho HR, Michaelis JE, Mutlu B. Investigating family perceptions and design preferences for an in-home robot. In: Proceedings of the Interaction Design and Children Conference. New York, USA: Association for Computing Machinery (ACM); 2020. pp. 229-242

[19] Levinson L, Gomez R, Šabanović S. Bow ties & colorful eyes: Centering youth designs of social robots. In: 2024 33rd IEEE International Conference on Robot and Human Interactive Communication (ROMAN). IEEE; 2024. pp. 1667-1674

[20] Kanda T, Sato R, Saiwaki N, Ishiguro H. A two-month field trial in an elementary school for long-term human–robot interaction. IEEE Transactions on Robotics. 2007;**23**(5):962-971

[21] O'Brien H. Theoretical perspectives on user engagement. In: O'Brien H, Cairns P, editors. Why Engagement Matters. Cham: Springer; 2016

[22] O'Brien HL, Cairns P, Hall M. A practical approach to measuring user engagement with the refined user engagement scale (UES) and new UES short form. International Journal of Human-Computer Studies. 2018;**112**: 28-39

[23] Papakostas GA, Sidiropoulos GK, Lytridis C, Bazinas C, Kaburlasos VG, Kourampa E, et al. Estimating children engagement interacting with robots in special education using machine learning. Mathematical Problems in Engineering. 2021;**1**:9955212

[24] Zhao Z, McEwen R. "Let's read a book together" a long-term study on the usage of pre-school children with their home companion robot. In: Proceedings of the 2022 ACM/IEEE International Conference on Human-Robot Interaction. New York, USA: Association for Computing Machinery (ACM); 2022. pp. 24-32

[25] Michaelis JE, Mutlu B. Reading socially: Transforming the in-home reading experience with a learning-companion robot. Science robotics. 2018; **3**(21):eaat5999

[26] Alimardani M, van den Braak S, Jouen AL, Matsunaka R, Hiraki K. Assessment of engagement and learning during child-robot interaction using EEG signals. In: Social Robotics: 13th International Conference, ICSR 2021, Singapore, Singapore, November 10–13,

2021, Proceedings 13. New York, USA: Springer International Publishing; 2021. pp. 671-682

[27] Dewan M, Murshed M, Lin F. Engagement detection in online learning: A review. Smart Learning Environments. 2019;**6**(1):1-20

[28] Levinson L, Kaufman Z, Karim A, Huang A, Gomez R, Sabanovic S. The nose knows: Using thermal imaging to approximate children's engagement with robots. In: Companion of the 2024 ACM/IEEE International Conference on Human-Robot Interaction. New York, USA: Association for Computing Machinery (ACM); 2024. pp. 669-673

[29] Sorrentino A, Fiorini L, Cavallo F. From the definition to the automatic assessment of engagement in human–robot interaction: A systematic review. International Journal of Social Robotics. 2024;**16**:1641-1663

[30] Pabba C, Kumar P. An intelligent system for monitoring students' engagement in large classroom teaching through facial expression recognition. Expert Systems. 2022;**39**(1):e12839

[31] Charisi V, Davison D, Reidsma D, Evers V. Evaluation methods for user-centered child-robot interaction. In: 25th IEEE International Symposium on Robot and Human Interactive Communication (RO-MAN). New York, USA; IEEE; 2016. pp. 545-550

[32] Wierzbicka A. Reading human faces: Emotion components and universal semantics. Pragmatics & Cognition. 1993;**1**(1):1-23

[33] Potamianos A, Narayanan S, Lee S. Automatic speech recognition for children. In: Fifth European Conference on Speech Communication and Technology. Grenoble, France: International Speech Communication Association (ISCA); 1997

[34] Russell M, D'Arcy S. Challenges for Computer Recognition of children's Speech. Workshop on Speech and Language Technology in Education. Grenoble, France: International Speech Communication Association (ISCA); 2007

[35] Liao H, Pundak G, Siohan O, Carroll M, Coccaro N, Jiang QM et al. Large Vocabulary Automatic Speech Recognition for Children. New York, USA; IEEE; 2015

[36] Gray SS, Willett D, Lu J, Pinto J, Maergner P, Bodenstab N. Child automatic speech recognition for US English: Child interaction with living-room-electronic-devices. WOCCI. 2014:21-26

[37] Van Straten CL, Peter J, Kühne R. Child-robot relationship formation: A narrative review of empirical research. International Journal of Social Robotics. 2020;**12**(2):325-344

[38] Beneteau E, Richards OK, Zhang M, Kientz JA, Yip J, Hiniker A. Communication breakdowns between families and Alexa. In: Proceedings of CHI Conference on Human Factors in Computing Systems. New York, USA: Association for Computing Machinery (ACM); 2019. pp. 1-13

[39] Xu Y, Warschauer M. Exploring young children's engagement in joint reading with a conversational agent. In: Proceedings of the Interaction Design and Children Conference. New York, USA: Association for Computing Machinery (ACM); 2020. pp. 216-228

[40] Zhao J, Wang G, Dally C, Slovak P, Edbrooke-Childs J, Van Kleek M, et al. I make up a silly name' understanding Children's perception of privacy risks

online. In: Proceedings of the 2019 CHI Conference on Human Factors in Computing Systems. New York, USA: Association for Computing Machinery (ACM); 2019. pp. 1-13

[41] Gavison R. Privacy and the limits of law. The Yale Law Journal. 1980;**89**(3): 421-471

[42] Nissenbaum H. Privacy as contextual integrity. Washington Law Review. 2004;**79**:119

[43] Lutz C, Schöttler M, Hoffmann CP. The privacy implications of social robots: Scoping review and expert interviews. Mobile Media & Communication. 2019;**7**(3):412-434

[44] Petronio S. Communication privacy management theory: Understanding families. In: Engaging Theories in Family Communication. Abingdon, UK: Routledge; 2017. pp. 87-97

[45] Petronio S, Child JT. Conceptualization and operationalization: Utility of communication privacy management theory. Current Opinion in Psychology. 2020;**31**:76-82

[46] Miltgen CL, Peyrat-Guillard D. Cultural and generational influences on privacy concerns: A qualitative study in seven European countries. European Journal of Information Systems. 2014; **23**(2):103-125

[47] Lutz C, Tamó-Larrieux A. The robot privacy paradox: Understanding how privacy concerns shape intentions to use social robots. Human-Machine Communication. 2020;**1**:87-111

[48] Charisi V, Gomez E, Mier G, Merino L, Gomez R. Child-robot collaborative problem-solving and the importance of child's voluntary interaction: A developmental perspective. Front Robot AI. 2020;**7**:15

[49] Leyzberg D, Ramachandran A, Scassellati B. The effect of personalization in longer-term robot tutoring. ACM Transactions on Human-Robot Interaction (THRI). 2018;**7**(3): 1-19

[50] Kory-Westlund JM, Breazeal C. A long-term study of young children's rapport, social emulation, and language learning with a peer-like robot playmate in preschool. Frontiers in Robotics and AI. 2019;**6**:81

[51] Pan Y, Hirokawa M, Suzuki K. Measuring K-degree facial interaction between robot and children with autism spectrum disorders. In: 24th IEEE International Symposium on Robot and Human Interactive Communication (RO-MAN). New York, USA; IEEE; 2015. pp. 48-53

[52] Esposito R, Fracasso F, Limosani R, Onofrio GD, Sancarlo D, Cortellessa G, et al. Engagement during interaction with assistive robots. International Journal of Social Robotics. 2018;**10**: 333-345

[53] Yun WH, Lee D, Park C, Kim J, Kim J. Automatic recognition of children engagement from facial video using convolutional neural networks. IEEE Transactions on Affective Computing. 2020;**11**(4):696-707

[54] Whitehill J, Serpell Z, Lin YC, Foster A, Movellan JR. The faces of engagement: Automatic recognition of student engagement from facial expressions. IEEE Transactions on Affective Computing. 2014;**5**(1):86-98

[55] Perry GT, Krippendorff K. On the reliability of identifying design moves in

protocol analysis. Design Studies. 2013; **34**(5):612-635

[56] Cohen J. A coefficient of agreement for nominal scales. Educational and Psychological Measurement. 1960;**20**(1): 37-46

[57] Cronbach LJ. Coefficient alpha and the internal structure of tests. Psychometrika. 1951;**16**(3):297-334

[58] Monarch RM. Human-in-the-Loop Machine Learning: Active Learning and Annotation for Human-Centered AI. New York, USA: Simon and Schuster; 2021

[59] Hemphill JF. Interpreting the magnitudes of correlation coefficients. The American Psychologist. 2003;**58**(1): 78-80

[60] Larson S, Mahendran A, Peper JJ, Clarke C, Lee A, Hill P, Kummerfeld JK, Leach K, Laurenzano MA, Tang L et al. An evaluation dataset for intent classification and out-of-scope prediction. arXiv preprint arXiv: 1909.02027. 2019

[61] Ljunglöf P, Larsson S, Mühlenbock KH, Thunberg G. TRIK: A talking and drawing robot for children with communication disabilities. In: Proceedings of the 17th Nordic Conference of Computational Linguistics (NODALIDA 2009). Uppsala, Sweden: Northern European Association for Language Technology (NEALT); 2009. pp. 275-278

[62] Li J. Recent advances in end-to-end automatic speech recognition. Asia-Pacific Signal and Information Processing Association. 2022;**11**(1):e8

[63] Adipat S, Laksana K, Busayanon K, Asawasowan A, Adipat B. Engaging students in the learning process with game-based learning: The fundamental concepts. International Journal of Technology in Education (IJTE). 2021; **4**(3):542-552

[64] Axelsson A, Skantze G. Multimodal user feedback during adaptive robot-human presentations. Frontiers in Computer Science. 2022;**3**:55-68

[65] Javed H. Personalizing mixed initiative dance interactions with a socially-aware robot (doctoral dissertation, The George Washington University)

[66] Vachharajani P. Enhancing social media personalization: dynamic user profile embeddings and multimodal contextual analysis using transformer models. arXiv preprint arXiv: 2407.07925. 2024

[67] Anderson DR, Bryant J, Wilder A, Santomero A, Williams M, Crawley AM. Researching blue's clues: Viewing behavior and impact. Media Psychology. 2000;**2**(2):179-194

[68] Komatsubara T, Shiomi M, Kanda T, Ishiguro H, Hagita N. Can a social robot help children's understanding of science in classrooms? In: Proceedings of the Second International Conference on Human-Agent Interaction. New York, USA: Association for Computing Machinery (ACM); 2014. pp. 83-90

[69] Pop CA, Simut RE, Pintea S, Saldien J, Rusu AS, Vanderfaeillie J, et al. Social robots vs. computer display: Does the way social stories are delivered make a difference for their effectiveness on ASD children? Journal of Educational Computing Research. 2013;**49**(3): 381-401

[70] Bainbridge WA, Hart JW, Kim ES, Scassellati B. The benefits of interactions with physically present robots over

video-displayed agents. International Journal of Social Robotics. 2011;**3**(1): 41-52

[71] Alvarez-Benito G, Suarez C, Marquez C. Hello I am a robot and my name is Curro: Design, development and evaluation of a robot solution for linguistic, cognitive and physical rehabilitation of children after cancer treatment. PanCare Conference. 2019

[72] Neugnot-Cerioli M, Laurenty OM. The future of child development in the AI era. Cross-disciplinary perspectives between AI and child development experts. arXiv preprint arXiv: 2405.19275. 2024

Chapter 4

Perspective Chapter: Social Awareness in HRI

Marcos Ribeiro Pereira Barretto and Vera Pereira-Barretto

Abstract

Increasingly, robots are becoming part of daily life: devices such as vacuum cleaners or self-driving cars are examples of robots interacting with humans, which necessitates an understanding of their social roles. This chapter explores the general requirements for human-robot interaction (HRI) in cases where robots directly engage with humans, proposing that they should be conceptualized as social robots. We identify core components for fostering social awareness in robots: morphology, dialog, effective communication, navigation, individuality, personality, privacy, and ethics. While some of these requirements are currently considered in robot design, they are often addressed without adequately accounting for the social environment in which the robot will operate. Beyond these core components, it is essential to evaluate a robot's functionality by taking its social role into account. Doing so will necessitate the incorporation of additional sensory systems and the establishment of behavioral rules to align with its intended social context.

Keywords: social robots, social awareness, social requirements in HRI, privacy in HRI, ethics in HRI

1. Introduction

More and more, robots interact with humans in daily life. Possibly, even an ordinary person interacts with a robotic vacuum cleaner, such as Roomba. In many stores, robots are being used as information panels, helping customers to find products or to answer questions about them, such as Pepper. In some cities, as in Los Angeles and San Francisco, robots are delivering goods, such as those from Serve Robotics. Or transporting people, like Waymo and other companies. Robotic assistants like ElliQ are being installed in nursing homes and retirement houses, helping to keep elderly people mentally active. Robotic toys such as AIBO and MISA are toys that keep children entertained.

In factories, robots are not kept behind fences anymore. A revised ISO10218 [1] standard is about to be published in 2025, bringing a necessary review since IMRs (industrial mobile robots) and collaborative robots are more and more frequent on the shop floor and in warehouses, working in an environment close to humans.

Some pivotal works discuss sociable robots in general, such as Fong et al. [2], Breazeal [3, 4], Mahdi et al. [5], and Leite et al. [6]. Other works discuss applications, such as education [7–9] or health care [10, 11]. These works were fundamental to help organize the list of general requirements discussed here.

The key point of this chapter is to reframe how we think about robots, and to add social awareness as an underlying requirement in all specific aspects. In fact, not only interactions with humans but also humans of all ages and needs (autism, impaired, blind, etc.), which imposes distinct requirements on the social behavior of robots, but also other sentient beings such as dogs and cats, which are present in daily life in our houses and streets. We discuss the general implications of social awareness in morphology, dialog, effective communication, individuality, personality, navigation, privacy, and ethics as fundamental aspects in robotics affected by social awareness. Also, we discuss briefly the impacts of social awareness in functionality, since it varies strongly from one robot application to another: impacts on social awareness for a vacuum cleaner robot are quite distinct from those for an assistant robot.

2. Social robots

The work of Fong et al. [2], though not recent, remains a cornerstone in the field of social robotics. Drawing upon the foundational definition by Dautenhahn and Billard, cited by Fong et al. [2], social robots are described as "embodied agents that are part of a heterogeneous group: a society of robots or humans. They are able to recognize each other and engage in social interactions, they possess histories (perceive and interpret the world in terms of their own experience), and they explicitly communicate with and learn from each other." Kirby et al. [12] says, "social robots are designed to interact with people in human-centric terms and to operate in human environments alongside people. Many social robots are humanoid or animal-like in form, although this does not have to be the case. A unifying characteristic is that social robots engage people in an interpersonal manner, communicating and coordinating their behavior with humans through verbal, nonverbal, or affective modalities". These illustrative definitions, among others, agree on the following characteristics:

- *Physical embodiment*, i.e., a social robot has a physical body;
- *Social skills*, i.e., a social robot interacts with humans and other sentient beings such as animals, following the social rules relevant to its role;
- *Autonomy*, i.e., a social robot makes decisions by itself.

Fong et al. [2] also reference Breazeal when categorizing social robots into four primary classes:

- *Function-oriented robots*: These robots are primarily designed to perform specific tasks or functions with some level of social interaction. Their main objective is practical utility, such as providing companionship or assisting in daily tasks.
    - *Example:* Robotic vacuums, like Roomba, fall into this category. But until now, they exhibited little social understanding.

- *Companion robots*: These robots are designed to provide social interaction and emotional support, mimicking social behavior and emotions. They are often employed to support humans emotionally, offering comfort and interaction.
    - Example: Paro, a therapeutic robot for elderly care, and Pepper, a humanoid robot for customer service, are notable examples.
- *Interactive robots*: Designed for engagement, these robots respond to human gestures, speech, or actions, often used in educational or entertainment contexts.
    - Example: Jibo, a social robot for family interaction, and Moxie, designed for educational purposes, exemplify this category.
- *Socially-aware robots*: These robots engage proactively with humans to satisfy internal social aims, such as drives or emotions. They require sophisticated models of social cognition.
    - Example: Kismet, developed at MIT, demonstrates internal states guiding its reactions.

The important aspect of a classification of social robots, being this or any other as those in Mavridis [13] and Bunt et al. [14] is that, clearly, social robots differ significantly from conventional industrial robots, teleoperated robots, and AGVs/AMRs and also from existing robots such as Roomba or Pepper, therefore bringing the need to consider social awareness as requirement in HRI design of robots interacting with humans.

3. Morphology

Morphology influences social interaction by shaping expectations: all humans judge based on appearance. The first look: that is the primary social awareness impact. For instance, a dog-like robot will elicit different human reactions compared to an anthropomorphic robot. However, a human-like appearance may not always be desirable due to the "uncanny valley" effect conceptualized by Mori [15] shown in **Figure 1**.

Figure 1 illustrates how familiarity varies according to various types of artifacts. It displays familiarity with both moving and still entities. An industrial robot shows little familiarity when compared to a humanoid robot. A prosthetic hand is frequently weird, exemplifying the drop of familiarity. Mori's original picture, as in **Figure 1**, was later somewhat refuted, particularly because of humanoid robots, which do not always display familiarity. Works such as Berns and Ashok [16] and Yam et al. [17] tried to investigate which anthropomorphism aspects result in familiarity, adding or removing them as "humanizing" or "dehumanizing" robot appearances. The results are not conclusive but clearly illustrate the phenomenon.

Fong et al. identify several morphological types:

- Anthropomorphic: Resembling a human.
- Zoomorphic: Resembling an animal.

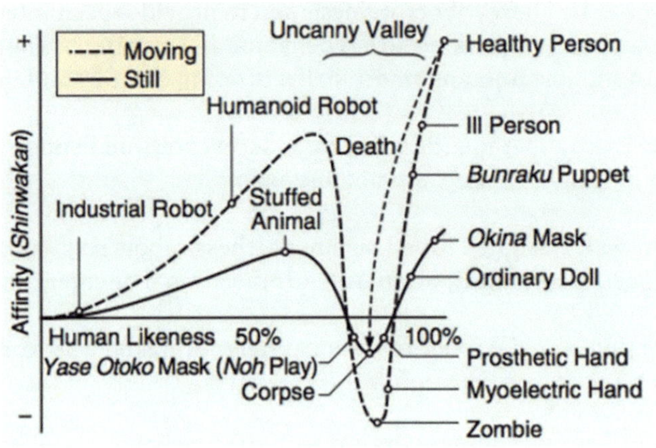

Figure 1.
The "uncanny valley" [15].

- Caricatured: Simplified or stereotypical embodiments.
- Functional: Prioritizing function over form.

Other classifications were proposed, such as Mahdi et al. [5], but the above is useful to classify most products discussed in this chapter: "functional" such as vacuum cleaners; caricatured, as most assistants such as ElliQ; anthropomorphic as Pepper, zoomorphic as Spot, Boston Dynamics dog. Morphology is the first drive of human expectation, a central aspect of social interaction.

4. Dialogue

Social interactions extend beyond simple commands and rely on conversational context. Here "dialog" is used to include all forms of communication, both verbal and nonverbal: voice, screens, touch screens, buttons, etc. Following Mavridis [13], some desired goals include, besides the "simple command":

- Multiple speech acts as in ISO24617-2 [14].
- Mixed-initiative dialog, as the robot should be able to initiate the dialog.
- Situated language and the symbol grounding problem.
- Affective interaction, as discussed deeper in Section 5.
- Motor correlates and nonverbal communication, also discussed in Section 5.
- Purposeful speech and planning, i.e., how much cheap chat is meaningful in HRI?

While human-robot communication takes many forms, Fong et al. [2] identify three primary types of communication media:

- *Low-level (pre-linguistic)*: Basic, nonverbal exchanges.
- *Nonverbal*: Gestures or other visual cues.
- *Natural language*: Conversational interactions enabled by advancements in large language models.

Among these, natural language has become increasingly feasible due to recent technological developments such as ChatGPT.

Dialog is absolutely central to social awareness.

5. Affective communication

Affective communication plays a critical role in human behavior. It includes:

- Verbal communication, as prosody conveying emotions.
- Nonverbal communication, particularly body gestures, is not only related to anthropomorphic robots but also to all other types of morphology: consider, for instance, your vacuum cleaner blinking an LED if it finds a harmful situation. Facial gestures are particularly relevant when conveying emotions. But body language in general is an important emotional display, using arms, hands, shoulders, and general posture. Touch should also be included in this category.

Emotional models in robotics are typically categorized into [18]:

- *Discrete approaches*: Using specific labels (e.g., happiness, sadness) to classify emotions.
- *Dimensional approaches*: Employing continuous values to represent emotional dimensions (e.g., arousal, valence).
- Componential theories, such as Scherer et al. [18], attempt to integrate discrete and dimensional approaches.

Affective communication in social robots is a subject with a large bibliography, such as Kirby et al. [12], Paterson [19], and Abdollahi et al. [20] to cite a few.

6. Navigation

Robot navigation in the presence of humans presents unique challenges in the field of navigation, as it necessitates the search for a socially acceptable path. The survey by Kruse et al. [21], although somewhat dated, remains a foundational reference on this topic.

A fundamental challenge in socially acceptable motion techniques is the accurate detection of individuals within the environment. Essential tasks for achieving this include pedestrian detection [22, 23], people tracking [24], and the recognition of human actions and activities [25, 26], among others. These steps are prerequisites for enabling socially compliant navigation.

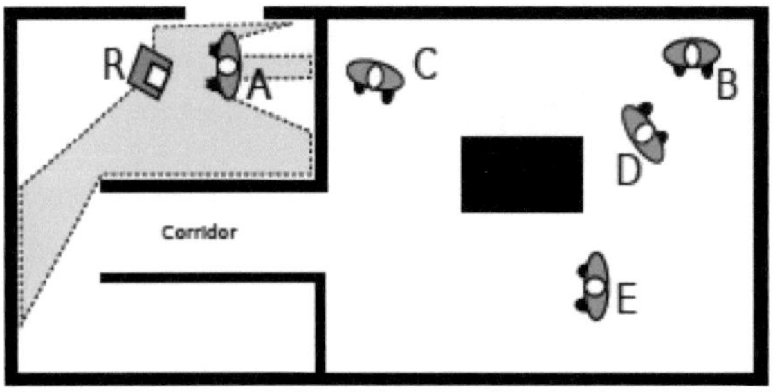

Figure 2.
Example scenario [21].

To provide a general understanding of the problem, Kruse et al. [21] discuss a scenario illustrated in **Figure 2**, where the robot is tasked with guiding Person A to Person B without disturbing other individuals in the environment.

Traditional trajectory planning techniques, such as obstacle avoidance, are employed to identify a feasible path. However, an added layer of complexity involves minimizing disturbances to Persons C, D, and E along the way. The robot must maintain sufficient space for Person A to follow it (or walk alongside it, depending on the specific task) while adhering to socially acceptable distances (approximately 1 meter, as suggested by Kruse) and maintaining an appropriate speed.

Kruse et al. identify key aspects of socially acceptable navigation:

1. *Comfort:* The absence of annoyance or stress for humans interacting with robots.

2. *Naturalness*: The extent to which the robot's low-level behavior patterns resemble those of humans.

3. *Sociability*: Adherence to explicit high-level cultural conventions.

In this context, "comfort" is considered a more nuanced concept than mere safety, as it encompasses the need for appropriate distancing. For human-to-human interactions, Hall cited by Kruse et al. [21] proposed the values summarized in **Table 1**.

However, further research is required to determine whether these values are equally applicable to HRI.

In addition to these social considerations, robots must also achieve conventional navigation goals, including task completion, energy efficiency, time efficiency, and ensuring safety.

Designation	Specification	Reserved for
Intimate distance	0–45 cm	Embracing, touching, whispering
Personal distance	45–120 cm	Friends
Social distance	1.2–3.6 m	Acquaintances and strangers
Public distance	>3.6 m	Public speaking

Table 1.
Social distancing [21].

7. Individuality

Fong et al. [2] introduce an important discussion regarding the distinction between individual and collective robots. Individual robots operate based on their unique experiences, whereas collective robots share knowledge within a network.

Research about robotic swarms, in general, relates to the resulting functionality, i.e., about the internal swarm behavior leading to some results, such as discussed in Duan et al. [27] and Bredeche and Fontbonne [28]. But collective robots can leverage learning, passing their experience to others, therefore helping to fulfill its function. As a simple example, a vacuum cleaner sharing his knowledge about a house may help in its replacement or help others in the same building to understand the house topology. In sections X and Y, we discuss some consequences in privacy and ethics of knowledge sharing.

An individual robot may learn from his experience, acquiring specific behavior based on these unique experiences. Still, in the simple example of a vacuum cleaner, it can understand where it's frequently dirtier and adapt its behavior to this. This concept ties into the notion of individuality. A social robot can exhibit individuality and even be recognized as an "electronic person," a term proposed by the European Parliament's Committee on Legal Affairs in a draft report on civil law rules for robotics [29]. This term envisions a legal status for sophisticated autonomous robots, granting them "specific rights and obligations, including that of making good any damage they may cause," and applying electronic personality in cases where robots make autonomous decisions or interact independently with third parties. Robot rights are derived from the legal discussion about animal rights and inspired the so-called "machine question" [30].

8. Personality

Personality significantly influences social interaction [31, 32]. Should a robot exhibit a distinct personality?

Consider the *Star Wars* franchise robots since they are exemplary models of social awareness, particularly R2-D2 and C-3PO.

R2-D2 assumes the role of a "mechanical technician" (referred to as an "astromech droid" in the *Star Wars* universe), adept at repairing machinery, interacting with systems, and responding to human commands. Additionally, it acts proactively, often anticipating human needs. Although R2-D2 operates under the instructions of a master, it occasionally circumvents legal constraints in service of its master's objectives, raising intriguing ethical considerations. R2-D2 does not speak in human language but understands it, communicating instead through "beeps" that humans interpret as a unique linguistic system. Its actions reflect courage, as it undertakes critical and dangerous tasks without hesitation. However, these behaviors are purely mechanical responses, devoid of human emotion. R2-D2 exemplifies loyalty, frequently risking its existence for its master and adhering to Asimov's Laws of Robotics [33]. Its understated heroism is marked by humility, as it seeks neither recognition nor praise, embodying the ideal functionality of a machine.

C-3PO, self-described as a "protocol robot," boasts the ability to communicate in over six million languages and comprehends a vast array of cultures, including their customs, traditions, etiquette, and ceremonial practices. Its primary function is to facilitate interaction among humans and other beings by providing translation and ensuring cultural appropriateness. In contrast to R2-D2, C-3PO features an anthropomorphic design, enhancing its relatability to humans. Notably, it exhibits a highly anxious demeanor, often fixating on minor details or potential dangers with repeated exclamations of "We're doomed!" It frequently highlights risks and expresses discomfort in unpredictable or chaotic situations. Despite its critical contributions, C-3PO tends to underestimate its capabilities, viewing its responsibilities as burdensome. The robot is verbose, often sharing excessive or tangential information, earning it the epithet "mindless philosopher" from Princess Leia. Its approach to problem-solving emphasizes logic and practicality, though it often defaults to pessimistic assumptions. Moreover, its literal thinking limits its ability to grasp sarcasm, adding to its endearing yet occasionally exasperating personality.

The richly developed personalities of R2-D2 and C-3PO often lead viewers to momentarily overlook their mechanical nature. These characters illustrate the potential for robots to engage humans on a social level while raising thought-provoking questions about the ethical and functional dimensions of advanced robotics.

9. Privacy

Privacy has become a critical concern in contemporary society. Personal data is continuously collected through various means such as GPS, cameras, smart devices, on-demand television, and more. Companies, and occasionally governments, utilize this data to uncover habits and tailor marketing strategies. To address these issues, many countries are enacting laws to protect data privacy. Robots, however, introduce additional complexity to this challenge as they access novel forms of data. For instance, home appliances like vacuum cleaners and ovens can collect information that traditional devices cannot capture.

To achieve optimal performance, robots must gather what can be termed "intimacy data," a category of personal information that extends beyond conventional data privacy considerations. This introduces ethical challenges, particularly with the emergence of robot swarming, where robots or humans share data among themselves. Imagine a scenario where your vacuum cleaner informs your neighbor about breadcrumbs under your bed or shares intimate photos without consent. Such possibilities highlight the privacy concerns tied to robotic data sharing.

These concerns raise important questions: Do privacy risks deter humans from adopting social robots? Is there a "privacy paradox" in which the benefits of social robots are weighed against fears of privacy loss?

Lutz and Tamo-Larrieux [34] conducted a study involving approximately 500 U.S. citizens aged 18–74. Using the model depicted in **Figure 2**, they explored the relationship between robot use intention and factors such as trust, privacy concerns, perceived benefits, scientific interest, and social influence. Additionally, the study examined how social influence impacts these factors. The experimental factors and their relationship are represented in **Figure 3**.

Their findings regarding physical privacy revealed:

- Trusting beliefs and privacy concerns had no significant effect on robot use intention, leading to the rejection of hypotheses H1 and H2.
- Perceived benefits positively influenced the intention to use social robots, supporting H3.

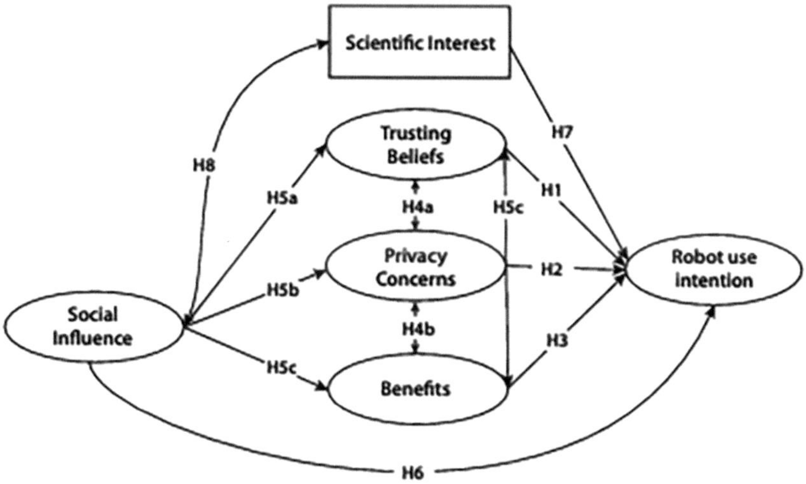

Figure 3.
Experimental representation according to Lutz and Tamo-Larrieux [34].

- Physical privacy concerns, trusting beliefs, and perceived benefits were significantly correlated and aligned with expectations, supporting H4.

- While physical privacy concerns were unaffected by social influence, social influence positively impacted trusting beliefs and perceived benefits, partially supporting H5.

- Social influence had a significant positive effect on robot use intention, supporting H6.

- Scientific interest did not significantly affect robot use intention, rejecting H7, but it was positively influenced by social influence, supporting H8.

The study also examined the effects of these factors on institutional informational privacy (concerning data usage by companies and governments) and social informational privacy (related to hacking and data breaches). Respondents expressed minimal concern about physical privacy but were significantly more worried about institutional privacy—specifically, data protection by manufacturers. There was moderate concern about malicious uses of social robots by other users, such as stalking or hacking. Overall, while respondents displayed moderate privacy concerns about social robots, other studies show notable apprehension about whether smart speakers like Alexa, Siri, or Google Home adequately safeguard privacy [35].

10. Ethics

Ethics form the foundation of all social interactions, serving as the guiding principles by which individuals and entities navigate complex relationships and dilemmas. This importance extends to the realm of robotics, where ethical considerations are critical in ensuring that technology aligns with human values. Discussions of robot ethics often invoke the "trolley problem" or "crash problem" in the context of self-driving cars [36]. For example, if a robot must choose between hitting a minivan with five passengers or a roadster with one person, it confronts a profound moral uncertainty. Here, the robot is compelled to make a life-altering decision, embodying the ethical dilemmas intrinsic to its programming.

However, moral uncertainty is not confined to vehicles. Consider agricultural robots operating in environments populated by animals. These systems must address ethical considerations involving sentient beings. For instance, should a robotic harvester prioritize the safety of a turtle crossing its path? Such scenarios underscore the need for ethical frameworks that guide robots in balancing operational efficiency with the preservation of sentient life.

Social robots raise unique ethical questions due to their direct interactions with humans. Empathy is a key attribute for these robots. For instance, personal assistant robots may face ethical dilemmas about prioritizing emotional well-being over truthfulness. Is it morally acceptable for a robot to lie to an elderly user by saying, "Your son called," when he did not? Such decisions involve weighing the benefits of emotional comfort against the intrinsic value of honesty.

Nursing robots, anticipated as essential in aging societies, present another layer of ethical complexity. Delegating decisions about people's care and well-being to algorithms raises significant concerns. Can an algorithm adequately consider the nuances of human dignity, autonomy, and emotional needs? Furthermore, elderly users often anthropomorphize their robotic companions, developing deep emotional attachments. This phenomenon, observed since the advent of ELIZA and continuing today with advanced conversational agents like ChatGPT, highlights ethical concerns about fostering dependency or escapism through prolonged interactions with robots.

The issue of deception by robots is another pressing ethical question. Should robots deceive humans through behavior or speech? Isaac and Bridewell [37] suggest that robots might need to employ "white lies" to better meet human expectations and maintain trust. However, such deception risks eroding the moral fabric of human-robot relationships and potentially manipulating users in ways that undermine their autonomy.

The ethical landscape becomes even more intricate when considering sex robots and military robots ("warbots"). These applications challenge societal norms and values in profound ways. For example, the on-demand series *Westworld* depicts an amusement park where robots enable the fulfillment of any human desire without consequence. This fictional scenario prompts critical reflection on the ethical implications of using robots to satisfy desires that might be harmful or morally questionable if directed toward humans. Such narratives force us to consider the boundaries of acceptable robot behavior and the societal impacts of normalizing certain actions through robotic intermediaries.

The integration of social robots into human life necessitates a robust ethical framework. This framework must address the moral uncertainty inherent in robotic decision-making, the balance between empathy and truthfulness, the risks of anthropomorphism, and the implications of deception. As robots become increasingly autonomous and entwined with human society, the ethical questions they raise will only grow more complex, requiring open and transparent discussions and probably new laws, under careful consideration to ensure that technology serves humanity in an equitable and just manner.

11. Social awareness applied to functionality

Social awareness affects functionality in specific ways for each robot application. To grasp this impact, imagine a vacuum cleaner equipped with social awareness. This capability could entail the following:

- It would refrain from cleaning if the baby sleeps in the room or if the user is watching a movie, demonstrating an understanding of human activities and preferences.

- It would recognize obstacles such as food dropped by a baby or pet waste, avoiding actions that could exacerbate messes instead of resolving them. (Personal anecdote: my robotic vacuum cleaner once spread dog urine across the bedroom, misidentifying it as a typical liquid.)

- It would communicate effectively with its user, potentially employing voice interaction for greater accessibility.

- It would provide relevant information about its operation, such as recommending more frequent activation based on observed needs.

- It would respect privacy by not sharing sensitive data with other devices, such as disclosing the types of debris found in the home to a neighbor's vacuum cleaner (assuming inter-device communication capabilities).

Such a vacuum cleaner would require contextual understanding and appropriate behavioral responses, executing or refraining from tasks based on situational demands. It would need to demonstrate empathy, effectively manage unexpected situations—possibly seeking human input—and communicate in a manner that aligns with human interaction norms (e.g., voice or visual feedback rather than buttons or complex interfaces). Moreover, it would need to adhere to ethical principles and maintain user privacy. While affectivity might not be essential, an understanding of emotional states would be crucial for effective human-robot interaction.

12. Conclusion

In this chapter, we postulate that robots with strong interaction with humans should be thought of as social robots.

A brief introduction to social awareness was presented, discussing its core components: morphology, dialog, effective communication, navigation, individuality, personality, privacy, and ethics. Even in well-developed fields such as navigation, taking social awareness into consideration brings new challenges.

But robots, as considered in this text, are not "general machines": they are built to perform a specific task, such as a vacuum cleaner. So, it's necessary to analyze each application to understand its social role and determine the characteristics to apply. A simple and partial example of a hypothetical vacuum cleaner with social awareness was present.

Social awareness represents perhaps the ultimate level of human-robot interaction, going far beyond interfacing with voice, video processing, and other basic functions in robotics. It does not require AGI (artificial general intelligence); it can be built with today's technology. But it's expected AGI includes complete social awareness.

We marvel at the new parkour of biped robots, but the real challenge to build a new generation of robots working close to humans is to build social awareness.

Acknowledgements

The authors thank Minerva Institute and CPS-PUC-SP for their support on this research.

Author details

Marcos Ribeiro Pereira Barretto[1*] and Vera Pereira-Barretto[2]

1 Minerva Institute, São Paulo, Brazil

2 Sociosemiotics Research Center (CPS) of PUC-SP, São Paulo, Brazil

*Address all correspondence to: marcos.barretto@minerva.org.br

References

[1] ISO10218. Robotis-Safety Requirements. Switzerland: International Standards Organization; 2025

[2] Fong T, Nourbakhsh I, Dautenhahn K. A Survey of Socially Interactive Robots: Concepts, Design, Applications. Pittsburgh, PA, EUA: The Robotics Institute at CMU; 2002. CMU-RI-TR-02-29

[3] Breazeal CL. Designing Social Robots. Boston, MA, EUA: MIT Press; 2002

[4] Breazeal C, Dautenhahn K, Kanda T. Social robotics. In: Siciliano B, Khatib O, editors. Springer Handbook of Robotics. EUA; 2016

[5] Mahdi H, Akgun SA, Saleh S, Dautehnhahn K. A survey on the design and evolution of social robots: Past, present and future. Robotics and Autonomous Systems. 2022;**156**

[6] Leite I, Martinho C, Paiva A. Social robots for long-term interaction: A survey. International Journal of Social Robotics. 2013;**5**:291-303

[7] Belpaeme T, Kennedy J, Ramachandran A, Scassellati B, Tanaka F. Social robots for education: A review. Science Robotics. 2018;**3**(21)

[8] Belpaeme T, Tanaka F. Social Robots as Educators. Paris, France: OECD Digital Education Outlook; 2021

[9] Lampropoulos G. Social robots in education: Current trends and future perspectives. Information. 2025;**16**(29)

[10] Breazeal C. Social robots for health applications. In: 33rd Annual International Conference of the IEEE EMBS. Boston, MA, EUA: IEEE; 2011

[11] Ragno L, Borboni A, Vannetti F, Amici C, Cusano N. Application of social robots in healthcare: Review on characteristics, requirements, technical solutions. Sensors, Basel. 2023;**23**(15):6820

[12] Kirby R, Forlizzi J, Simmons R. Affective social robots. Robotis and Autonomous Systems. 2010;**58**:322-332

[13] Mavridis N. A review of verbal and non-verbal human-robot interactive communication. Robotics and Autonomous Systems. 2015;**63**:22-25

[14] Bunt H, Petukhova V, Gillmartin E, Pelachaud C, Fang A, Keizer S, et al. The ISO standard for dialogue act annotation, second edition. In: Proceedings of the 12th Conference on Language Resources and Evaluation. Marseille, France: European Language Resources Association; 2020

[15] Mori M, Mac Dorman K, Kageki N. The uncanny valley [from the field]. IEEE Robotics and Automation Magazine. 2012;**19**(2):98-100

[16] Berns K, Ashok A. "You scare me": The effects of humanoid robot appearance, emotion and interaction skill on uncanny valley phenomen. Actuators. 2024;**13**(10):419

[17] Yam KC, Bigman Y, Gray K. Reducing the uncanny valley by dehumanizing humanoid robots. Computers in Human Behavior. 2021;**125**

[18] Scherer KR, Schorr A, Johnstone T. Appraisal Processes in Emotion. Oxford, UK: Oxford University Press; 2001

[19] Paterson M. Social robots and the futures of affective touch. The Senses and Society. 2023;**18**(2):110-125

[20] Abdollahi H, Mahoor MH, Zandie R, Siewierski J, Qualls SH. Artificial emotional intelligence in socially assistive robots for older adults: A pilot study. IEEE Transactions On Affective Computing. 2023;**14**(3):2020-2032

[21] Kruse T, Pandey AK, Alami R, Kirsch A. Human-aware robot navigation: A survey. Robotics and Autonomous Systems. 2013;**61**(12):1726-1743

[22] Dollar P, Wojek C, Schiele B, Perona P. Pedestrian detection: An evaluation of the state of the art. IEEE Transactions on Pattern Analysis and Machine Intelligence. 2012;**34**(4):743-761

[23] Mahmoud A, El-Barkouky A, Graham J, Farag A. Pedestrian detection using mixed partial derivative based histogram of oriented gradients. In: IEEE International Conference on Image Processing (ICIP). Paris, France: IEEE; 2014. pp. 2334-2337

[24] Andriluka M, Roth S, Schiele B. People-tracking-by-detection and people-detection-by-tracking. In: IEEE Conference on Computer Vision and Pattern Recognition. Anchorage, AK, EUA: IEEE; 2008. pp. 1-8

[25] Hu N, Englebienne G, Lou Z, Krose B. Learning latent structure for activity recognition. In: IEEE International Conference on Robotics and Automation (ICRA). Hong Kong, China: IEEE; 2014. pp. 1048-1053

[26] Song Y, Morency L, Davis R. Multi-view latent variable discriminative models for action recognition. In: IEEE Conference on Computer Vision and Pattern Recognition (CVPR). Providence, RI, EUA: IEEE; 2012. pp. 2120-2127

[27] Duan H, Huo M, Fan Y. From animal collective behavior to swarm robotics collaboration. National Science Review. 2023;**10**

[28] Bredeche N, Fontbonne N. Social learning in swarm robotics. Philosophical Transactions on the Royal Society B. 2022;**377**(1843)

[29] Delvaux M. Available from: http://www.europarl.europa.eu/sides/getDoc.do?pubRef=-//EP//NONSGML%2BCOMPARL%2BPE-582.443%2B01%2BDOC%2BPDF%2BV0//EN [Accessed December 21, 2024]

[30] Gunkel D. The Machine Question: Critical Perspectives on AI, Robotics and Ethics. Boston, EUA: MIT Press; 2012

[31] Whitaker S, Rogers Y, Petrovskaya E, Zhuang H. Designing personas for expressive robots: Personality in the new breed of moving, speaking and colorful social home robots. ACM Transactions on Human-Robot Interaction. 2021;**10**(1):1-25

[32] Sorrentino A, Khalid O, Coviello L, Cavallo F, Fiorini L. Modelling human-like robot personalities as a key to foster socially aware navigation. In: 30th IEEE International Conference on Robot and Human Interactive Communication. Vancouver, Canada: IEEE; 2021

[33] Asimov I. Available from: https://www.wikiwand.com/en/articles/Three_Laws_of_Robotics [Accessed: December 21, 2024]

[34] Lutz C, Tamo-Larrieux A. The robot privacy paradox: Understanding how privacy concerns shape intentions of use social robots. Human-Machine Communication. 2020;**1**:87-111

[35] Chung H, Iorga M, Voas J, Alexa LS. Can I trust you? Computer. 2017;**50**(9):100-104

[36] Lin P, Abney K, Jenkins R. Robot Ethics 2.0. Oxford, UK: Oxford University Press; 2017

[37] Isaac AMC, Bridewell W. White lies on silver tongues: Why robots need to deceive (and how). In: Lin P, Jenkins R, Abney K, editors. Robot Ethics 2.0. Oxford University Press; 2017

Section 2

Mobile Robots

Chapter 5

Perspective Chapter: Advanced Environment Modelling Techniques for Mobile Manipulators

Noelia Fernandez, Gonzalo Espinoza, Alberto Mendez, Adrian Prados, Alicia Mora and Ramon Barber

Abstract

In recent years, the capabilities of mobile robots to interact with humans and their environment have been enhanced by emerging technological advances. The improvement in the quantity and quality of information from robot sensors, along with increased computational capacities, has enabled the development of new algorithms to model the human environment in which the robot moves and interacts. This chapter will describe modelling techniques for scenes and objects that compose the environment, with the aim of generating an internal representation of the robot's surroundings so it can navigate and perform manipulation tasks in shared environments with humans. Recent advances in environment modelling driven by deep learning and machine learning techniques, using sensor information, will be presented. Additionally, new trends in the generation of simulated environments, their elements, and synthetic databases will be studied. These trends aim to leverage advances in artificial intelligence to expand the quantity and variety of available data, facilitating the process of learning and understanding the environment.

Keywords: mobile manipulator robots, environment mapping, object modelling, data augmentation, simulation platforms

1. Introduction

Recent technological advancements have significantly enhanced the capabilities of mobile robots. Improvements in sensor technology have increased the volume and precision of the information that robots can capture from their surroundings. Alongside these advances, the development of more powerful hardware has enabled the processing of this vast amount of data, extending the robots' abilities to acquire knowledge about their environment and make informed decisions. These achievements have made human-robot interaction (HRI) increasingly prominent, enabling robots and humans to collaboratively perform tasks within shared workspaces. This progress not only facilitates safer and more efficient operations but also highlights the importance of seamless integration between robotic systems and human partners in various real-world applications.

However, many challenges in robotics remain unsolved. Robots should count on the capability of moving in a safe and efficient manner in diverse complex environments, for which they must be capable of perceiving, interpreting, and representing their environment in a precise and robust way. This becomes particularly critical when interacting with humans, as ensuring user safety must not come at the expense of task performance. Just as humans create internal representations of their surroundings in their brains, robots rely on models to plan navigation and manipulation tasks. Consequently, it is crucial to optimise resources and develop new techniques to leverage available information for reconstructing the environment within the robot. Effective modelling and learning are essential for robots to perceive, interpret, and interact with their environments accurately. The complexity of mapping lies in the necessity of integrating information coming from diverse sensors and continuously updating the map in response to new observations. Successfully integrating this information will allow robots to perform tasks that involve interaction with both the environment and humans in a natural and seamless manner.

This chapter will delve into environment modelling, exploring various approaches and recent advancements in the field, with the aim of creating scenarios that facilitate effective interaction with humans. It will cover advances in environment modelling for navigation, advances in object modelling for manipulation and grasping, techniques for data augmentation and generation to enhance robot capabilities, and finally, the use of simulations as platforms for testing modelling algorithms. Each section will highlight key developments and methodologies, as well as current challenges and recent trends, providing a comprehensive overview of the current state of environment modelling in robotics.

2. Environment modelling for robot navigation

Autonomous mobile robot navigation is a core field of study in modern robotics, and it has been constantly evolving in recent years. This discipline aims to provide robots with the capability of moving in a safe and efficient manner in diverse complex environments. In order to achieve this goal, robots must be capable of perceiving, interpreting, and representing their environment in a precise and robust way. This process is known as mapping, and it is crucial not only for autonomous planning and decision-making processes but also for enabling seamless and safe interaction with humans in shared spaces. Effective mapping allows robots to avoid obstacles, optimise their behaviours, and adapt to changing environments while safeguarding user safety and ensuring collaboration.

Mapping implies building internal representations of the outside world with which robots can localise and move. The complexity of mapping lies in the necessity of integrating information coming from diverse sensors and continuously updating the map in response to new observations. Including human presence as a dynamic element in the map adds an additional layer of complexity, requiring models that can anticipate and respond to human behaviours to ensure fluid and natural interactions.

The applicability of the resulting environment model created during the mapping process depends on the abstraction level at which information is captured and represented [1]. There are several approaches to this process, among which geometric, topological, and semantic mapping stand out.

Geometric mapping focuses on detailed representations, including physical characteristics of the environment. Topological mapping simplifies the space representation by creating nodes that include relevant information and connecting them using links. Semantic mapping adds a higher level of understanding by including the meaning and functionalities of objects and regions in the environment. Each of these approaches has its own advantages and challenges, which can be combined in multiple cases to find the most efficient solution for autonomous robot navigation. Additionally, the presence of users in the environment should be addressed. This section reviews the main characteristics of each method and highlights recent trends in each field.

2.1 Geometric mapping

Geometric mapping focuses on representing precise information and physical characteristics of the environment. This approach has been the basis for robot navigation since its beginnings, so it has been developed and refined over several decades. Geometric mapping captures environmental information using sensors like LiDARs, cameras, and sonars. Captured data is processed to create detailed representations in 2D or 3D, describing the location and shape of the elements around the robot so that it can differentiate between traversable and non-traversable regions.

One of the most challenging problems in geometric mapping is generating a map while locating the robot in it, which is known as the SLAM (Simultaneous Localisation and Mapping) problem. In order to know where to place any sensor information on a map, it is necessary to know where the robot is located. At the same time, sensor information must be compared to a map to update it. SLAM solves this consistency by matching consecutive scans and re-calculating the current robot position inside the map during scan registration.

Traditionally, SLAM has been used to create occupancy grid maps, which represent the environment as a grid in which each cell indicates the probability of that coordinate being occupied. One of the first works on geometric mapping [2] highlights the required components for recovering spatial information from sensor data:

- *Coordinate transformations*, which allow to concatenate sensor measurements taken from different positions into a unified world model. This is represented as a set of robot poses, both position and orientation, from which data is captured.
- *A spatial interpretation model*, adapted to each different kind of sensor, which translates the sensor data into a statement about which areas are occupied or empty.
- *A map updating model*, which composes views provided by the sensor into a single grid representation.

This idea has been preserved in modern applications. Additionally, further complementary implementations have been developed to take into account mapping errors caused by challenging properties of the environment. Mora et al. in their study [3] propose the detection of reflective surfaces by analysing laser scan intensity values in order to build a reflection-aware map for safe indoor robot navigation, as shown in

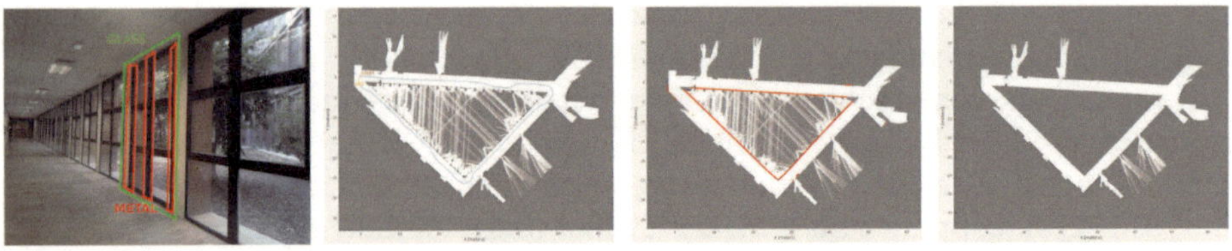

(a) Real Scenario (b) Original Map (c) Reflective Zones (d) Modified Map

Figure 1.
Trends in geometric mapping. A challenging real scenario (a) with glass and metallic elements is mapped (b) by detecting reflecting surfaces (c) and modifying the occupancy grid map (d) for its application in robust autonomous navigation.

Figure 1. The main novelty of the proposed method is the usage of a single sensor to gather environmental information and a single parameter to determine where reflective zones like glass walls are. The work of Thomas et al. [4] proposes the construction of a Spatiotemporal Occupancy Grid Map (SOGM) to consider dynamic elements of the environment. The method generates, predicts, and uses these maps to embed future information of dynamic scenes to enable lifelong learning for robots. In the work of Funk et al. [5], the mapping procedure is extended to create a 3D representation. It proposes an efficient system that leverages the concept of adaptive-resolution volumetric mapping, naturally integrating with the hierarchical decomposition of space in an octree data structure which allows for collision queries, as needed for robot motion planning.

2.2 Topological mapping

Topological mapping is an alternative approach to geometric mapping that simplifies the representation of the environment by means of graphs that represent relationships between several places or points of interest without the need of capturing their exact geometry. Instead of focusing on geometric precision, topological mapping focuses on creating nodes that represent significant locations in the environment and creating and maintaining connectivity among them. These models allow robots to understand the structure of the environment in terms of connectivity, which is particularly advantageous when navigating environments shared with humans, as it allows robots to focus on functional areas and pathways that are more relevant to human activities.

There are two main sources of information that are used for building topological maps: sensors and a previously built geometric map. In the first case, the topological map is incrementally built online. The robot uses sensors such as cameras to automatically detect significant characteristics in the environment such as corners or specific objects, which are represented by a node. Then, links are created to represent the connectivity between them. It is essential to have a loop closure methodology to identify if a newly detected node has been previously detected to update the topological graph accordingly and avoid mapping the same information multiple times. In the second case, a geometric map like an occupancy grid map is analysed to extract relevant information like narrow passages, which typically correspond to doors, to segment the environment and extract the topological map. An example of this procedure is shown in **Figure 2**. In this case, the loop closure problem is not relevant as it is assumed that it has been previously solved by the geometric mapping algorithm.

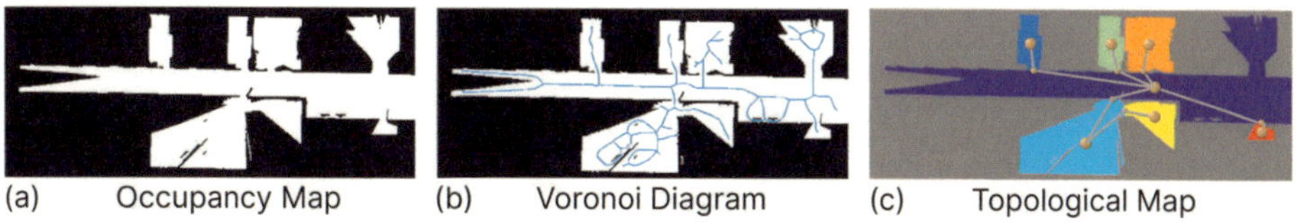

Figure 2.
Trends in topological cartography. A gridded occupancy map (a) is segmented by analysing the free regions. The mapped area is segmented into regions corresponding to rooms and corridors by means of a Voronoi diagram (b) and a topological map is extracted (c), where the nodes correspond to navigable areas and doors.

However, there are other limitations, such as fully relying on a single map, which may be noisy.

Topological maps are particularly useful for indoor environments like homes or office buildings, where the structure and layout of sites are highly relevant. Robots can effectively navigate in these locations using a topological map that identifies rooms, corridors, doors, and any other point of interest without the need of specifically identifying the exact shape or precise location of these spaces. This facilitates interpreting scenes and significantly reduces the computational complexity required for navigation.

This mapping methodology is more efficient in terms of storage and processing requirements with respect to geometric models, as it is a simplified representation of the environment. Moreover, it simplifies tasks like planning for environments that are clearly structured. However, topological mapping is less precise, so tasks like robot localisation can be challenging. Another main challenge is adapting graphs to dynamic or unstructured environments, where the shape and exact location of elements are constantly changing or difficult to estimate.

Recent works attempt to address these challenges. In the work of Liao et al. [6], a novel approach to generate a scene graph from RGB-D data is proposed. A topological map is built to represent rooms in an indoor environment, where connections between them indicate robot behaviours for navigating. Additionally, it includes a scene graph for each room in which spatial connection relations like 'next to', 'on', or 'under' are defined between detected objects, expanding the modelled knowledge of the environment. This enriched representation supports robots in better understanding and interacting with human-occupied spaces. Similarly, Rosinol et al. in their work [7] extend the definition of topological maps to create dynamic scene graphs (DSGs), which allow to represent dynamic scenes with moving agents and include information about feasible actions to support planning and decision-making applications. It enables robots to adapt their behaviour based on the dynamic changes in human environments, improving both safety and efficiency in shared spaces.

2.3 Semantic mapping

Semantic mapping is an advanced approach that goes beyond purely representing elements with geometric features or graphs. It incorporates a higher level of understanding including the meaning and functionality of objects and places in the environment. This methodology allows robots to not only perceive and understand the layout of the environment but also to understand the relevance of the elements around them and the feasibility of their use in the tasks that they must perform. This capacity is crucial for enabling robots to function effectively in human-centric environments, where understanding context and intent is essential for seamless interaction.

Semantic maps turn out as a helpful tool for representing indoor environments since they allow robots to understand the utilities of the objects in the scene. For instance, a domestic robot could identify the kitchen or a bedroom and the elements inside them like chairs, tables, or household appliances. By modelling these elements and their utilities, the robot could perform tasks like cleaning a room, delivering objects, or preparing food by matching the required tasks with the objects that need to be involved in them. This higher-level understanding is pivotal for collaborative scenarios where robots and humans share tasks, ensuring the robot can act intelligently and safely in dynamic environments.

This field of mapping in robotics is the most recent and the one that has been most driven by recent technological advances. Artificial Intelligence (AI) plays a major role in the generation of these maps, as it provides the necessary tools for data processing and applying learning models for extracting semantic information.

The main source of information for semantic maps is computer vision approaches, which are trained to recognise and classify objects. They can be applied to images and also to point clouds, thus allowing the extraction of information in two and three dimensions, assigning semantic labels to objects that describe their functionality and relevance in the environment. This information leads to one of the most common semantic mapping approaches, which is object-based maps. **Figure 3** shows an example of an object-based map extracted from a reconstructed room. These maps identify and classify objects in the environment in order to allow robots to understand the layout of the scene and to use this information for planning and executing specific tasks in an efficient manner. A representative example can be found in the work of Martins et al. [8], where object-level information is used in several applications like assistive robotics or visual navigation. The robot needs an internal semantic representation of the scene and a real-time detector for grounding the internal representations with the real world so that models can be updated according to changes.

Recent trends and advances in artificial intelligence, especially deep learning, have broadened the capabilities of robots to acquire and model semantic knowledge of the environment beyond computer vision techniques.

One of the most recent trends is the integration of transformers-based natural language models in robotic modelling applications. This tool can help with the comprehension of textual descriptions of the environment, like verbal instructions. Huang et al. in their work [9] propose the creation of visual language maps, a spatial map

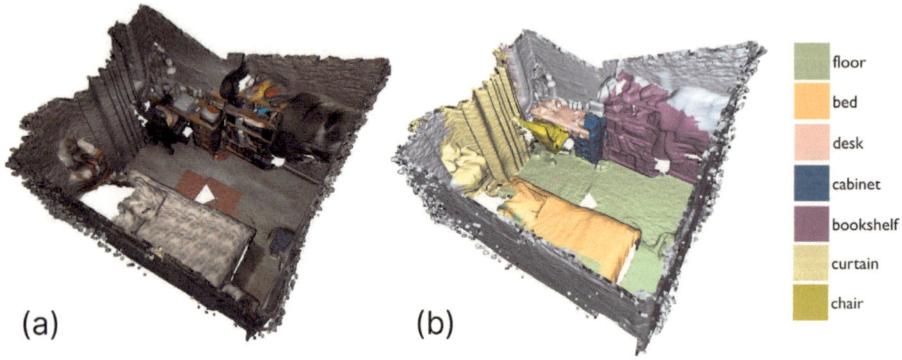

Figure 3.
Trends in semantic mapping. Reconstructed indoor scene using visual SLAM (a) and the corresponding semantic map extracted by labelling the elements on the environment (b).

representation that fuses pre-trained visual language features that match images to natural language descriptions of object goals with a 3D reconstruction of the physical world to create a semantic representation.

Another relevant trend is transfer learning, which allows robots to use previously obtained knowledge and apply it to new environments or tasks. By using pre-trained models on large-scale databases, robots can adapt their semantic understanding capacities to specific scenes with minimal additional training. The work presented by Al-Halah et al. [10] proposes a unified approach to semantic navigation using a modular transfer learning model that enables zero-shot experience learning, solving tasks without receiving sky task-specific interactive training.

Semantic maps provide robots with the advantage of understanding their surroundings and interacting with them in an efficient way by precisely adapting their skills to new situations. By providing a deeper understanding of the environment, robots can interact more naturally and effectively with their surroundings and with human users, increasing the number of complex tasks that can be performed. However, the construction of semantic maps can be more complex and requires a larger amount of data and computational resources compared to other mapping approaches. In addition, the accuracy and robustness of the semantic map may be affected by the diversity of objects and situations to be recognised and understood. The continuous evolution in this field promises to significantly improve the autonomous capabilities of robots and expand their applications in a wide range of environments and tasks.

2.4 Modelling for human-aware navigation

Navigating human-populated spaces is a complex task for robots, requiring solutions to various engineering and social challenges. Research in robotics and HRI has led to significant developments in addressing these issues. Social navigation involves not only ensuring the safety of both the robot and humans but also respecting social norms such as personal space and social rules in confined areas. Key topics in social navigation include trajectory prediction and planning, as well as understanding and responding to human behaviour.

Trajectory prediction and planning can be approached by either decoupling or integrating prediction and planning processes. Decoupled methods, like dynamic obstacle avoidance, predict human movements independently of the robot's actions, using techniques such as inverse reinforcement learning and Bayesian reasoning [11]. These methods treat humans as dynamic obstacles or consider social objectives for dynamic avoidance [12]. Integrated approaches, however, account for the mutual influence between robot movements and human behaviour. Cooperative Collision Avoidance (CCA) models [13] are developed to predict joint movements of all agents in shared environments, with some methods explicitly coupling prediction and planning, while others use interaction-aware objectives for decision-making.

Ensuring efficient and safe robot navigation around humans is crucial, but achieving socially competent behaviour requires considering higher-level social norms. Proxemics defines personal space and group formations which robots should avoid disrupting. The analysis of human gestures, enhanced by artificial intelligence, plays a significant role in this context [14]. Understanding human behaviour improves planning performance and fosters human comfort and trust in robots. Techniques like Gaussian Mixture Models dynamically model personal and interpersonal spaces to prevent robots from intruding into these zones, ensuring both safety and social appropriateness in human-robot interactions [15], as shown in **Figure 4**.

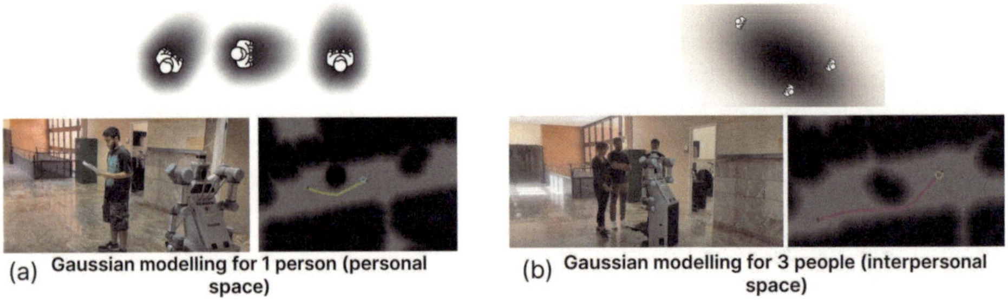

Figure 4.
Personal space modelling using Gaussian mixture models for navigation considering human space presented in the work of Mora et al. [15]. (a) Modelling of the personal space of a single user, (b) modelling of the interpersonal space between several users, the robot does not pass through so as not to disturb the users.

3. Advanced techniques for object modelling

As humans, we subconsciously build hierarchical structures of our environment. For example, a floor of a building is divided into rooms, which are further divided into workstations containing various objects. This structure helps us distinguish between entities with different shapes, sizes, physical properties, and functionalities, allowing us to group similar objects and generalise information about them. Similarly, robots need object models or internal representations of their environment, which enable them to understand and process the features of different objects, facilitating knowledge inference and decision-making.

Accurate object representations are therefore essential, especially in manipulation and grasping tasks, for the robot to predict the object's behaviour during interactions with humans and to plan an appropriate response to it. The need for better object modelling has been more recently highlighted by the development of assistant and social robots which often must adapt to the human environments in which they operate, which require robot manipulators, mobile or fixed, to possess a major degree of intelligence and autonomy to face the difficulties of uncertain environments. The following sections will address the newest techniques used for object modelling specially oriented towards manipulation and grasping, which are summarised in the scheme shown in **Figure 5**.

3.1 Techniques for 3D object modelling

Opposite to environment modelling for robot navigation, 3D representations of objects are imperative for manipulation and grasping tasks. Even in the case where manipulation tasks involve trajectories within a plane, 3D information remains essential for identifying an object's size, shape, or location and effectuating secure grasps. Thus, several 3D representation methods have been studied and analysed to exploit their particular advantages.

3.1.1 Types of 3D representation

One of the most important 3D representation techniques for objects is point cloud representation, which consists of representing the environment as collections of single points in a 3D space. Each point represents a measurement of 3D coordinates, sometimes complemented with colour or brightness of the surface at that particular

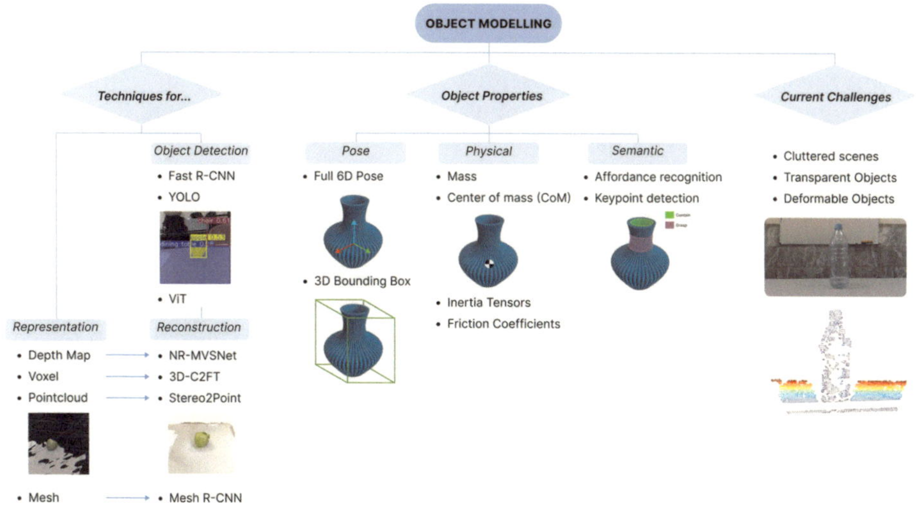

Figure 5.
Advanced techniques for object modelling. A visual scheme that includes techniques for modelling, methods for object properties estimation, and the current challenges of object modelling.

location [16]. These data can be acquired from different light emission-reception-based sensors such as LiDARs and stereo cameras which also allow to obtain RGB images of the scene. However, despite being accurate representations, point clouds are unstructured collections of data and require processing through noise filtering and outlier removal [17]. Once the point clouds are obtained, the object modelling is a computer vision problem divided into different steps, from object detection and recognition, followed by object segmentation, and finally object reconstruction. These topics are later covered in Section 3.1.2.

On the other hand, 3D meshes are particularly useful in grasping research. Meshes, composed of vertices, edges, and the surfaces they define, provide valuable information about an object's topology and surface normals, which are often used to evaluate and rank different grasp poses of a robot hand around an object. For instance, Du et al. in their work [18] use mesh representations to optimise an objective function aimed at transferring human grasp demonstrations to robotic grippers. This function enhances the alignment between the hand mesh model and the object's mesh surface normals at contact points and penalises mesh interpenetration by quantifying the number of vertices inside the opposite mesh. Similarly, Lundell et al. in their work [19] use an estimated object mesh to refine grasp poses by adjusting contact points between the gripper and object vertices, while also avoiding interpenetration.

Lastly, depending on the application or the implementation, there has been interest in other object representations, such as voxel and computer-aided design (CAD). Voxel representation divides the point cloud into a regular grid, where each voxel is an occupied grid cell, which may contain further feature information about the points they correspond to [20]. This type of representation allows more efficient management of data and feature extraction, being well suited for convolutional neural networks (CNN) applications [20]. However, the discretisation of space often leads to a loss of detail in the geometry and surface of the object [21]. Moreover, 3D mesh models are also used for simulating robot-object interactions and evaluating grasping and manipulation methods based on planning and machine learning approaches. A popular dataset developed for this purpose is the YCB Object Set presented in the work of Calli et al. [22], which includes texture-mapped 3D mesh models and RGB-D images for a total of 77 distinct objects. Similarly, CAD object model sets, such as the well-known ShapeNet repository presented in the work of Chang et al. [23], have been developed, motivated by the benefits of using CAD models to represent the object surfaces with a high level of detail.

3.1.2 Techniques for object detection and reconstruction

Object detection and reconstruction are critical processes in robotics, as they enable precise manipulation of the objects in the environment and safe human-robot interaction. Object detection involves identifying and localising objects in an image or video stream. Traditional techniques used hand-crafted feature extractors, but these models were slow, inaccurate, and did not work well on unknown datasets. However, with the increase in computing power and deep learning, a variety of techniques have emerged. These approaches fall into three categories: two-stage, single-stage, and transformer-based detectors [24].

Two-stage detectors first generate region proposals and then classify and refine these proposals for accurate object detection, as in Fast R-CNN and Mask R-CNN. On the other hand, single-stage techniques, such as You Only Look Once (YOLO), directly predict object classes and bounding boxes for the entire image in a single evaluation step. Finally, transformer-based detectors, such as Vision Transformers (ViT), employ attention mechanisms to directly model global relationships, predicting objects and their locations in parallel [24].

Between object detection and object reconstruction, an intermediate step is the fusion of sensor information to extract the 3D shape of an object as can be seen in **Figure 6**, this is the proposal of Mora et al. [25]. However, the limitation of this method is that the shape of the object is not complete as a result of occlusions. For this reason, for accurate object manipulation, object reconstruction is performed using different techniques.

Object reconstruction, on the other hand, aims to recreate a three-dimensional model of an object from two-dimensional images. This reconstruction can be done from a single image or using multiple views depending on the method. In multi-view 3D reconstruction, two main approaches stand out: image feature fusion and shape feature fusion. Image feature fusion combines sets of extracted features into a unified representation during encoding, then decodes to form the composite model. In contrast, shape feature fusion fuses shape features after the encoding and decoding process, focusing on preliminary models that are combined for the final result. The latter is more accurate with limited training data, capturing the distinctive topology, while image feature fusion excels on large datasets. Also, within these ideas, techniques can be based on different methods of representation. In the same way, within

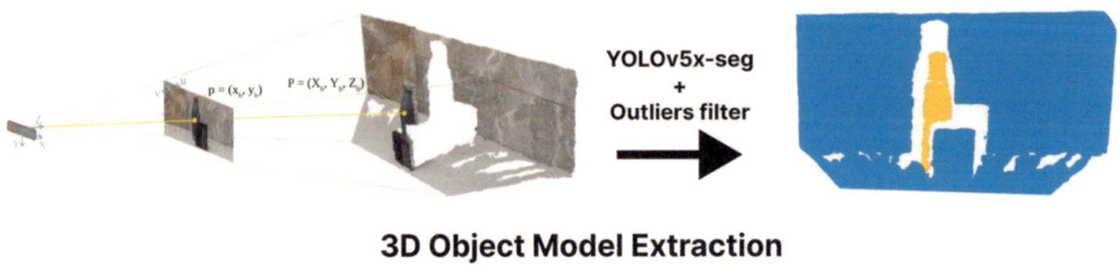

Figure 6.
3D Object modelling using sensor fusion in a single view configuration.

these ideas, techniques can be based on different kinds of 3D representation such as depth maps, voxels, point clouds, and meshes [26].

The depth map-based techniques use depth information from different perspectives to generate the 3D geometrical shape of the object, while voxel-based techniques excel for simpler objects with clear geometry. On the other hand, point cloud-based models directly manipulate point cloud data and utilise deep learning methods for its reconstruction. And mesh-based techniques accurately depict surface structures, reduce memory usage, enhance reconstruction efficiency, and effectively describe local surface details. Within all these techniques, the performance of the following approaches is remarkable: for depth map-based reconstruction, the NR-MVSNet model; for voxel-based reconstruction, the 3D-C2FT model; for using point cloud data, the Stereo2Point model, and, lastly, for mesh-based reconstruction, the Mesh R-CNN, as Juhao et al. highlighted in their survey [26].

3.2 Methods for objects properties estimation

To interact with objects, robots, much like humans, must understand properties such as weight and location, obtained through their sensory systems, to effectively plan manipulation strategies. For successful manipulation and grasping, key object features to be modelled include geometric, physical, and semantic properties. Physical properties help determine the object's centre of mass, grasp forces, and friction between the robot and the object's surface. At a higher level of abstraction, semantic information enables the robot to understand an object's function. For instance, tools can be grasped in multiple ways, but only specific grasps will allow for their correct use.

3.2.1 Object pose

While an object's shape and size can be directly extracted using various 3D representation techniques, estimating the 6D pose of objects remains a challenging yet crucial task in robotics, especially for manipulation and grasping. In fact, object pose estimation is a vital step towards automating manipulation tasks in uncertain environments where object locations are unknown. Furthermore, precisely determining an object's position is essential to assess if it is reachable by the robot's arm, to generate robust grasp poses, and, in the case of ambidextrous robots, to determine the optimal arm for grasping the object.

According to Sahin et al. [27], the methods in the literature focus on the estimation of the object's pose as either 3D bounding boxes or full 6D poses. In the first case, 3D bounding boxes can be parameterised by their centre $\mathbf{x} = (x, y, z)$, size $\mathbf{d} = (d_w, d_h, d_l)$, and orientation in Euler angles $\Theta = (\theta_r, \theta_p, \theta_y)$. Depending on the application, the assumption of the alignment of the 3D bounding box with the gravity direction also serves to simplify the problem by reducing the searching space. Alternatively, the full 6D pose estimation approach seeks to determine the translation $\mathbf{x} = (x, y, z)$ and rotation $\Theta = (\theta_r, \theta_p, \theta_y)$ vectors of the object.

Several methods in the literature tackle both the detection and pose estimation problems simultaneously. For instance, Brachmann et al. in their work [28] use a decision forest based on an energy function that aims to predict both the object instance and its 6D pose. Hu et al. [29] use a different approach based on a CNN encoder-decoder architecture, which is split into two branches for estimating the object instance and its oriented 3D bounding box.

3.2.2 Physical properties

Physical properties are essential for predicting an object's behaviour during motion. Properties such as mass, centre of mass (CoM), inertia tensors, and friction coefficients provide critical information about an object's heaviness, weight distribution, resistance to motion change, and the relative motion between two contacting surfaces. For this reason, accurate estimation of these properties is crucial for robust manipulation task planning. However, as noted by Mavrakis and Stolkin [30], the lack of sensors for direct measurement of inertial parameters forces researchers to rely on indirect methods such as visual estimations, experimental interactions with the object, or measuring wrenches on robot joints. Although some of these methods can yield accurate estimations, the absence of direct measurement limits the full utilisation of inertial parameters.

To demonstrate the importance of physical properties, Feng et al. in their work [31] develop a regrasp planner, which corrects initial grasp poses using CoM estimations when slip is detected. Slippages occur due to the lack of knowledge about the object's weight distribution, which is then estimated by using tactile sensors, completing the visual information.

3.2.3 Semantic features

Semantic features enrich object properties by adding contextual and functional information. This improves the robot's ability to manipulate these objects by understanding their functionality and keypoints, drawing inspiration from how humans perceive and interact with their environment. Currently, there is a new trend powered by deep learning techniques known as Affordance recognition and Keypoint detection. This method allows the robot to extract these semantic features from an object that defines relevant points and guidance on the affordance and how to affect it.

Xu et al. in their work [32] proposed an affordance keypoint detection network that successfully detects the five affordance categories labelled (grasp, contain, cut, pound, and w-grasp) and its keypoints to estimate the grasp and the direction of manipulation for the action associated with each object.

3.3 Current challenges of object modelling

The integration of mobile robots into human environments has introduced new challenges in manipulation and grasping that remain unresolved. These challenges stem from the need to adapt robots to diverse and unpredictable human environments rather than modifying the environments to suit robots. Consequently, robots often encounter scenarios where objects are not ideally positioned or where the objects themselves present significant difficulties. For example, robots frequently face cluttered scenes where multiple objects are occluded, resulting in only partial visibility of the objects. Moreover, perception systems must adapt to handle transparent or deformable objects. This section presents several methods developed to address some of the most common and complex problems in this area.

Most of the methods discussed earlier assume that objects are rigid or nearly rigid, meaning the distance between any two points within the object remains constant, making their behaviour during motion easier to predict. However, many applications, such as medical robotics and social robotics, involve interacting with deformable objects like fabrics, rubbers, food items, plastics, organs, and papers, whose manipulation is far more complex. Siciliano and Ruggiero in their work [33] identify three main modelling approaches: mesh-based, mesh-free, and hybrid. Mesh-based approaches often use mass-spring models for planar objects, where the deformable object consists of discrete points connected by springs. Volumetric models are better represented by finite element methods (FEM), where each mesh element is governed by differential equations, and the object's behaviour is derived from solving these equations. Mesh-free approaches, on the other hand, use collections of free particles governed by physical properties, making them more suitable for modelling fluids and viscous materials.

Another significant challenge in manipulation and grasping in unknown environments is the perception of transparent objects, such as glasses, plastic containers, or objects with transparent parts. The difficulty in perceiving these objects arises because light passing through them is transmitted and refracted rather than evenly reflected (contrary to the Lambertian assumption). This makes it difficult for light-based sensors to capture accurate depth information. As detailed by Jiang et al. in their work [34], several datasets have been developed, including TransCut, GDD, ClearGrasp, ODD, and ClearPose, for transparent object segmentation, reconstruction, and pose estimation. These methods, which include deep learning, NeRF, and other multi-view and multimodal techniques, aim to address the perception challenges posed by transparent objects.

4. Data augmentation and generation for environment modelling

The methods for modelling described in the previous sections are not always capable of extracting sufficient data for learning and understanding the features of the observable environment. Currently, there is a new trend of techniques based on deep learning such as Latent Diffusion Models (LDM) to improve the capabilities of mobile robots in daily tasks, such as rearranging objects or sweeping the floor.

4.1 Augmentation methods

The modelling of the environment for robots is directly dependent on the available data. To apply the techniques described in the previous sections, there is a need for both 2D and 3D data. Sometimes, it can be difficult to have the large amount of data needed to apply these methods, so a practice called data augmentation is used, which involves increasing the quantity and diversity of the available information without collecting additional data. In this section, we will focus on the new trends for image and 3D augmentation.

4.1.1 Image augmentation

Collecting a balanced and large enough image dataset for recent Deep Learning models can be a very tedious task. One cause can be the time required for image classification and labelling. But usually, the most compelling reason for opting for data augmentation lies in the scarcity of data in certain applications, such as harvesting tasks or in indoor environments where, for privacy reasons, it is difficult to find images publicly. Image augmentation has been recognised as a suitable solution for this issue.

Image augmentation can be done by applying different techniques, which can be divided into three groups: model-free, model-based, and optimising policy-based. This section will focus on the first two as they are the most used for object detectors in mobile manipulators. Model-free algorithms, as the name suggests, are able to perform image augmentation without any model. Instead, model-based techniques need a pre-trained model to obtain the augmented images [35].

Model-free methods can use a single image or multiple images at once to generate the augmentation. With a single image, it is possible to apply geometrical transformations like translation or flipping, colour image processing such as jittering, or intensity transformations like blurring or adding noise. When using multiple images, the augmentation consists of adopting and fusing their features in a new augmented image, overlaying images, or doing copy and paste of objects between the images. These changes can be applied at a non-instance or instance level.

Model-based methods can affect the image under different conditions being unconditional, label conditional, and image conditional [35]. In this way, the novelty consists of doing data augmentation, adding new information to the image, or generating a new one via generative models [36]. Unconditional techniques use a pre-trained model to generate new images from scratch similar to the rest of the dataset. With label conditional techniques, there is one approach similar to copy and paste methods, where these generative models can include new objects with a specific label in the image. Finally, the image conditional techniques exploit the capacity of these models to generate new images, changing the style, light conditions, and other features using an image-to-image process. A detailed explanation of the generation process with these models can be found in Section 5.2.1.

4.1.2 3D augmentation

Augmenting 3D data poses a greater challenge due to the complexity of applying transformations to spatial data. Unlike 2D images, 3D data involves additional dimensions and geometric considerations, making the process of scaling, rotating, and translating more intricate. Each transformation must be carefully executed to maintain the integrity and accuracy of the 3D structures. Consequently, developing robust augmentation techniques for 3D data requires advanced algorithms and a deeper understanding of spatial relationships to ensure effective and realistic enhancements. These techniques can be applied to perform data augmentation, generation of synthetic data, trajectory and manipulation learning, and social navigation.

As discussed in the previous section, there are two types of approaches. On one hand, model-free proposals, such as the method proposed by Kar et al. [37], leverage the geometric information of the scene to achieve more realistic data augmentation. In this method, RGB information is combined with depth and in some cases, meshes, to generate changes in the scene such as modifying the point of view, adding motion blur, or altering lighting. Another approach is the Mix3D [38] technique, which involves combining two scenes into a new one. This helps deep learning models to better understand the object instances independently of their context scene.

On the other hand, there are many model-based approaches that may use modelling tools or generative models. Greff et al. [39] in their work present Kubric, a data generation pipeline for creating semi-realistic synthetic multi-object videos with physics. This method uses Blender as the rendering engine for rendering the scenes required in the scripts where you can adjust the data format (depth map, point cloud, or mesh), object movement, and textures. While Voleti et al. [40] in their work present a model able to generate a 3D object with a few images given.

4.2 Generation of synthetic data using generative models

As an alternative to data augmentation, a new trend driven by artificial intelligence is dataset generation. This approach involves collecting various types of data using generative models, whether images, videos, or even 3D object models, with the goal of training deep learning models to solve specific tasks. In this way, researchers can create large and varied datasets for applications where such data is difficult to access, thus improving the robustness and performance of deep learning algorithms in various applications.

4.2.1 Image generation

The advent of models capable of generating images from text has transformed the landscape of this field of research. A variety of models are capable of performing this task, including Diffusion Models (DMs), Generative Adversarial Networks (GANs), and the current trend, Latent Diffusion Models (LDMs). The latter have several advantages over more traditional methods, including enhanced computational efficiency, adaptability, and the generation of high-quality images.

LDMs are deep learning models that employ diffusion models and autoencoders in latent space to generate high-resolution images from a conditional input, which may be text or an image. The process by which they generate an image can be described as follows: the image is first compressed to the latent space, after which a probabilistic diffusion process is applied, which adds noise to generate variations. This process is then reversed iteratively in order to improve the image. Finally, a decoder is applied, which generates the final image from the latent representation [41].

The current state-of-the-art LDM is SDXL [36]. This model represents an improvement on its predecessor, StableDiffusion. The enhancements are achieved through the incorporation of a UNet-based backbone, which triples the number of parameters in comparison to the previous version; the introduction of additional conditioning stages; and the integration of a refining model in the final stage.

This model is a powerful AI tool that can be used to generate a dataset from scratch providing the desired objects and scenes as a prompt input. Also, it can be used to generate more images from an existing dataset with the image-to-image generation that allows to change the style of the image. Additionally, it can add new objects inside the image through the inpainting process. However, it should be noted that these datasets would not be perfect and need to be monitored due to the limitations of the model. Some of the problems with the model are the difficulty to represent in detail some structures such as human hands, the correct representation of light or textures, the confusion of input concepts, and the generation of readable text in the images [36].

4.2.2 Video generation

Based on LDMs, new models for video generation have recently appeared with a similar performance allowing text-to-video and image-to-video generation. Blattmann et al. [42] in their work present Stable Diffusion Video, a proposal that takes LDMs trained for image generation in combination with capable temporal order and finetuning on video datasets to obtain the ability to generate video. This model's ability to maintain temporal and spatial cohesion between frames also provides a foundation for 3D generation models.

Similar to the model discussed in the previous section, Stable Diffusion Video can generate datasets from scratch based on input prompts or create motion and style variations within existing datasets.

4.2.3 3D object generation

As previously stated, a video model can be transformed into a model capable of generating orbital videos around an object. This process, described by Voleti et al. [40], involves adapting an image-to-video model for multi-view and 3D generation of an object. In this work, the state-of-the-art method called SV3D is presented. This approach is particularly noteworthy given the difficulty in reconstructing a 3D model of an object from a single image in robotics.

For this purpose, the Stable Video Diffusion model is adapted by introducing an explicit camera pose conditioning and three main properties: pose-controllable, multi-view consistent, and generalisable. The process of reconstructing the 3D model of an object from a single image consists of two main stages [40]. First, the model performs the multi-view synthesis with its corresponding poses and time consistency. Then, the model has two possibilities to generate the 3D mesh using the orbital video generated as direct reconstruction targets (with a NeRF) or using Score Distillation Sampling (SDS).

This model can function as a 3D dataset generator using existing image datasets or, in conjunction with another LDM designed for image generation, create a dataset from scratch. A notable application in robotics involves reconstructing objects of interest within an environment to facilitate grasping and manipulation tasks, effectively addressing challenges such as occlusions and misshapen point clouds.

5. Modelling of the environment in simulation

Complementary to the techniques for augmenting real data from the environment, there has been an increased use of simulation environments to accelerate the process of data collection. Simulators are then essential for the design, planning, analysis, control, and decision-making processes in various development areas [43]. By enabling the replication of physical or theoretical systems, simulations facilitate exhaustive testing, highlight short-term difficulties, and evaluate results in a safe, rapid, and efficient manner. Since it is imperative to physically test robots before their implementation to ensure both environmental and self-integrity, simulations optimise solutions by quantifying uncertainty, detecting failures, and preventing undesired behaviours.

5.1 Simulation environments

In robotic systems, two types of simulations can be distinguished [44]: those based on tools for general systems and those based on specialised tools for robotic systems. Simulations for general systems include modules, libraries, or user interfaces that simplify and facilitate the construction of robotic environments. Some of the most well-known tools in this category are MATLAB/Simulink, Dymola/Modelica, 20-sim, and Mathematica. Specialised simulations for robotic systems cover one or more specific tasks in robotics, such as mechanical, kinematic, or dynamic design, and focus on mobile robots, industrial robots, manipulators, humanoids, and parallel mechanisms.

5.1.1 Importance of simulation in robotics HRI

Simulations are a critical interdisciplinary field used across domains such as engineering, computer science, economics, and social sciences, all of which intersect in the design and evaluation of human-robot interaction systems. All these areas are key and play a fundamental role in the continuous development of robotic systems. Some of its benefits are [45]:

- *Generation of large amounts of data explicitly and at low cost:* The recent increase in the use of machine learning has highlighted the need for large, validated training datasets. The simulation platform becomes a testing ground for creating models that learn from their errors and transfer their knowledge to real devices.

- *Improvement and minimisation of robot design cycle time:* The two most time-consuming stages in designing a robotic model are structural design and control. On the one hand, a model is created that modifies and improves its morphology through iterative loops, detecting failures until it adapts to the final requirements. On the other hand, the robot needs to be endowed with intelligence to carry out tasks. Simulation allows this to be done by reducing physical testing costs and increasing both software and hardware performance.

- *Accelerated, safe, and controlled virtual testing environments:* It is necessary to establish usage protocols for robots, and for this, system verification is carried out through the repetition of tasks in real time, eliminating human or hardware risks. Simulation offers collaborative multi-robot scenarios where they can interact based on their own local decision-making algorithms or include human interaction.

However, there are challenges or limitations that need to be mitigated. In the work of Collins et al. [46], a user survey identifies some issues. One of them is the gap between reality and simulation, as reality cannot be fully replicated. Additionally, there is complexity in simulator configuration due to the high degree of knowledge and time invested in model design. The lack of resources due to paid simulators, the difficult continuous integration of environmental elements to keep the model updated, and the high cost of computational resources are some examples.

5.1.2 Simulation tools

The options for commercial simulators are very varied. Advances and the perpetual change of technology favour the emergence of new simulators, while others fall into disuse. The application for which the simulation is intended and the degree of realism are factors to consider when selecting a simulation environment. The most notable ones for robotics are listed below. **Table 1** shows the different characteristics that can be found between the different simulators depending on the user's need, highlighting their features and the capabilities they offer users to replicate reality.

- *Chrono:* an open-source platform primarily used for off-road vehicle simulation. It is widely utilised by NASA, the US Army, and the US Navy, as it allows for the simulation of real vehicles with wheels and tracks, and offers easy interface configuration through Python.

Simulator	Chronos	Coppeliasim	Gazebo	Isaac	Unity	Webots
GPS	✓	✓	✓	✓	✓	✓
LiDAR	✓	✓	✓	✓	✓	✓
Tracks	✓	✓	✓	✓	✓	✓
Wheels	✓	✓	✓	✓	✓	✓
Omni wheels	✓	✓	✓	✓	✓	✓
Heightmap import	✓	✓	✓	✓	✓	✓
OpenDrive	×	×	×	×	×	✓
OpenStreetMap	×	×	×	×	×	✓
Pathplanning	×	✓	✓	✓	×	✓
ROS support	×	✓	✓	✓	×	✓
RGBD	✓	✓	✓	✓	✓	✓
Realistic rendering	✓	×	×	✓	✓	×

Table 1.
Comparison between popular simulators.

- *Coppeliasim:* formerly known as V-REP, is a closed-source multi-robot platform that includes multiple dynamic engines such as Mujoco, Newton Dynamics, Vortex Studio, and Bullet. The variety of simulated robots ranges from manipulators and robotic arms to manufacturing/automation robots and ground robots. Scripts can be configured in various languages (Lua, Python, or Java), though they offer limited and reduced sensor support.

- *Gazebo:* possibly the most widely used simulation platform, thanks to its realistic preconfigured scenes, sensors, and controllers. It supports deep learning from data, and although the physics engine is quite simple, users can extend it since it is open-source. It is used in various fields of robotics, and its integration with ROS is direct.

- *Isaac:* the latest simulator created by NVIDIA. It uses the Unreal Engine physics engine and implements HRI. It features very realistic predefined scenes, multiple sensors, and cameras. However, it is not open-source, and updates and compatibility with real devices depend exclusively on the company.

- *Unreal engine and unity:* the most versatile platforms for video game development, as they offer physics support, advanced graphics, scripting sequences, and collision detectors, among other features. Despite not being open-source, they are used due to their high degree of realism.

- *Webots:* one of the most popular open-source platforms for robots, used for over two decades. It offers realistic indoor scenes, as well as support for sensors, drones, and ground robots. It is mainly used for its ability to export code to real platforms.

5.1.3 Creation of simulation scenarios

It is crucial to distinguish between simulating a robot and the environment with which it will interact. Being able to recreate what the robot will perceive and its interaction with the environment ensures that the model's behaviour matches reality. Most scenarios of interest in robotics can be simulated and are based on four types of elements [47]: rigid bodies, flexible (deformable) bodies, fluids, and terrains or granular material.

- *Rigid bodies:* Robots and environmental elements are simulated by composing these elements, calculating the position, velocity, and acceleration of each rigid body at each moment in time. Sometimes, geometry is not important, and only the mass and moment of inertia of the element are needed. However, geometry is crucial in two cases: when the body collides with other bodies or when it is desired to visualise how that body moves over time.

- *Flexible bodies:* These are elements with a certain degree of elasticity through the generation of stresses and deformations. There can be elements with small deformations, large deformations, displacements or rotations, and non-linear materials.

- *Fluids or granular material:* If the robot moves or manipulates a fluid, its movement is calculated based on the Navier-Stokes momentum equilibrium equations and mass conservation equations. This allows identifying the location of volume change over time.

- *Terrain:* Mobile robots operating indoors glide over smooth surfaces with a certain degree of friction. However, mobile robots operating outdoors deal with uneven terrains that can sometimes be deformable. If the terrain is uneven but rigid, simulations are relatively simple as they only consider the effects of friction, contact, and impact. Conversely, if the terrain is deformable, the approaches used are of three types: semi-empirical, continuous formulations, and discrete representations. All of these take into account the material's viscosity and plasticity and the uniaxial pressure-sinkage of the robot on the terrain.

5.2 Use of physical laws

The use of physical laws in simulation allows for the prediction of the behaviour of a dynamic system of interest and its evolution over time. In robotics, the most relevant laws are the conservation of mass and the conservation of momentum. Additionally, the laws of optics are used when sensors or reaction forces of joints are needed, which often require non-trivial calculations with complex recursive formulations. All these laws are formulated through mathematical equations associated with dynamic systems, and due to the high level of complexity, specialised numerical analysis is required to provide an approximate solution. The software code that poses and solves these equations is called the dynamic engine.

Complex robotic systems such as mobile robots, drones, or humanoids consist of bodies with closed loops and a minimal set of generalised coordinates relative to a reference system. When establishing the equations of motion for each body, it is necessary to include kinematic constraints that may be linear or non-linear depending on the requirements. These constraints replicate the limitations that exist in reality and transfer them to the simulator to ensure the results obtained are as realistic as possible. Additionally, there is uncertainty associated with the parameters of the physical robot such as masses, moments of inertia, spring stiffness, actuator parameters, joint flexibility, or elastic properties of the material. Uncertainty is managed in three different ways [47]:

- *Sensitivity analysis:* Although there are many ways to evaluate it, the most commonly used in robotics is variance, which relatively classifies the importance of a model parameter based on the expected reduction in the model output.

- *Model calibration:* Since some values cannot be measured directly, model calibration provides values such as the coefficient of friction and Young's modulus, which are used in calculation processes that generate distributions later modified with real readings.

- *Monte Carlo analysis:* This is the least costly method and is based on understanding how the uncertainty value is reflected in the system output. It requires performing many simulations, hence the need for speed in the process.

5.3 Robot-environment interaction

The responses and solutions provided by simulators achieve a high degree of realism when they faithfully simulate robot-environment interaction. To achieve this, it is crucial to consider a series of factors:

- *Sensor simulation* [47]: Sensors such as cameras, LiDAR, IMU, and GPS are fundamental for environmental perception. In simulation, these sensors generate synthetic information that accurately replicates the conditions the robot would encounter in the real world. This includes simulating image capture, depth detection, orientation measurement, and global positioning. The accuracy and performance of the data generated by these virtual sensors are critical for the robot's autonomous decision-making and action planning.

- *Actuator simulation* [47]: Actuators, such as motors controlling the movement of wheels and robotic arms, are simulated to model how they physically respond to control signals. In simulation, how control signals translate into precise forces and movements is mapped, which is essential for optimising the robot's performance in various tasks. Actuator simulation allows for the adjustment and optimisation of control algorithms before physical implementation, thus reducing the time and costs associated with experimental development.

- *Human-robot interaction* [45]: Simulating the interaction between humans and robots is essential for developing autonomous and semi-autonomous systems that can effectively collaborate with people. This involves modelling not only the robot's physical capabilities but also its emotional responses and communication behaviours. HRI simulation allows for the adjustment of psycho-social factors and the evaluation of user interface effectiveness, facilitating the creation of safe and collaborative work environments.

- *Obstacles* [47]: Both static and dynamic obstacles are simulated to understand how they affect the robot's trajectory planning and navigation. In simulation, the robot's interactions with walls, furniture, people, and other moving objects are modelled. This includes simulating collisions, friction, and contact forces, providing the robot with a detailed understanding of its environment and allowing real-time adjustments to avoid accidents and optimise operational efficiency.

The rapid development of digitisation, the Internet of Things (IoT), and Industry 4.0 has led to the emergence of the digital twin concept [48]. This concept, which originated in 2003, is defined as a precise real-time replica of a physical system and encompasses three essential components: the physical component, the virtual component, and the connection that facilitates the exchange of information between them. Digital twins enable the optimisation of control and operations in real time by analysing continuous incoming data, making immediate adjustments, and leveraging predictive capabilities [49, 50]. In addition to vitalising the environment and modelling systems, they are used to analyse trajectories and consumption times, and predict behaviours with the goal of optimising processes.

6. Conclusions

The capabilities of mobile robots have seen substantial enhancement due to recent technological advancements. Improvements in sensor technology have provided robots with more accurate and extensive data, while advancements in computational hardware have enabled the efficient processing of this information. These developments have paved the way for sophisticated algorithms that model the environment, allowing robots to navigate and perform tasks with greater precision and autonomy.

Despite these advancements, numerous challenges in robotics remain, particularly in the realm of HRI. Effective navigation and manipulation in diverse and complex environments require robots to not only perceive, interpret, and accurately represent their surroundings but also to understand and adapt to human intentions and behaviours in real time. This necessitates advanced modelling techniques that integrate data from multiple sensors, dynamically update the environment map, and contextualise information for interaction. Developing these models is crucial for robots to perform safely and efficiently, mimicking the human ability to create internal representations of the world while enabling meaningful and intuitive interactions with humans.

This chapter has explored various approaches and recent advancements in environment and object modelling which have highlighted key issues to address in future research. From the perspective of *robot navigation*, classical techniques for environment mapping have been extensively studied and developed, leading to robust algorithms that generate reliable solutions. However, semantic map generation algorithms still present certain limitations in their implementation today. Many of these techniques require a large amount of data to be processed simultaneously as they need to perform a class assignment process based on the structural characteristics of elements in the environment. Additionally, semantic applications involve certain types of relationships between objects as well as their functionality, which often complicates their implementation for specific real-time tasks. To address these limitations, many algorithms tend to rely on artificial intelligence capable of extracting these relationships, which in turn demands high memory processing for the proper implementation of models. Regarding the modelling of humans in the navigation environment, the most important factor to consider for the future of this research line is the predictability of human movements and actions. This is a critical factor, as it is not only necessary to model users but also to anticipate their actions to allow robots to proactively act, thus avoiding undesirable interactions such as collisions.

Regarding *object modelling*, while research on object detection and pose estimation has shown remarkable success, object reconstruction remains constrained by the amount of information the visual sensors can provide and the management of data. Consequently, 3D object generation models are predicted to play an important role in object reconstruction from single image or image sequences. These will help to overcome challenges like occlusions through imaginative processes in the same way as how humans mentally visualise objects, and add semantic meaning in order to derive or predict the properties of objects prior to interaction. Furthermore, modelling is frequently approached at the object level to examine the manipulation of the entire object. Modelling at the part level would then be useful for tackling the challenge of using tools in manipulation tasks.

Concerning *data augmentation and generation*, augmentation methods have proven invaluable tools for improving the performance and robustness of state-of-the-art object detectors, pose estimators, and reconstruction models. However, future models that aim to achieve a deeper understanding of the environment and integrate rich semantic information will require significantly larger datasets. This demand poses a major challenge in the field of robotics research. To address this problem, recent advancements in deep learning, particularly in generative models, offer a promising solution to this data bottleneck by facilitating the creation of vast amounts of synthetic data. However, these techniques are still in their infancy and require substantial improvement to fully meet this challenge. Achieving this goal will enable the construction of comprehensive world model frameworks capable of integrating perceived, predicted, and previously known information about the environment. Such models bring robots closer to understanding and interacting with their surroundings as humans do.

Lastly, modelling in *simulation* enables the ability to work remotely on the creation of algorithms for robot interaction with the environment and users without the need for a physical robot, accelerating the testing process. The implementation of realistic simulation environments that integrate physical, dynamic, and cognitive elements allows for results that more accurately reflect real-world scenarios. By incorporating precise simulations of rigid and flexible bodies, advanced sensors, and human-robot interaction models, the developed systems gain an enhanced ability to interpret objects in their surroundings and respond effectively to human actions. This alignment between simulated and real worlds not only optimises robot design and control but also improves their performance in collaborative and autonomous tasks, ensuring safe, efficient, and adaptive human-robot interaction across various scenarios. However, it is important to emphasise that a deployment process in the real world must always be carried out, as slight differences between simulation and reality may exist. This is a complex and delicate process that is currently under investigation, aiming to create models that are as faithful as possible to real environments to minimise the gap between both of them.

In conclusion, advancements in environment modelling not only enhance the technical capabilities of robots but also lay the groundwork for more natural, secure, and effective human-robot collaboration. The future of HRI will require an interdisciplinary approach that integrates technological innovation, cognitive understanding, and adaptive systems to build truly collaborative robotic systems.

Acknowledgements

This work was supported by the project Advanced Mobile dual-arm manipulator for Elderly People Attendance (AMME) (PID2022-139227OB-I00), funded by the Spanish Ministerio de Ciencia e Innovacion.

Conflict of interest

The authors declare no conflict of interest.

Author details

Noelia Fernandez†, Gonzalo Espinoza†, Alberto Mendez†, Adrian Prados†, Alicia Mora† and Ramon Barber*†
Universidad Carlos III de Madrid, Leganés, Spain

*Address all correspondence to: rbarber@ing.uc3m.es

† All authors contributed equally.

References

[1] Alatise MB, Hancke GP. A review on challenges of autonomous mobile robot and sensor fusion methods. IEEE Access. 2020;**8**:39830-39846

[2] Matthies L, Elfes A. Integration of sonar and stereo range data using a grid-based representation. In: Proceedings 1988 IEEE International Conference on Robotics and Automation. Philadelphia, PA, USA: IEEE; 1988. pp. 727-733

[3] Mora A, Prados A, González P, Moreno L, Barber R. Intensity-based identification of reflective surfaces for occupancy grid map modification. IEEE Access. 2023;**11**:23517-23530

[4] Thomas H, de Saint Aurin MG, Zhang J, Barfoot TD. Learning spatiotemporal occupancy grid maps for lifelong navigation in dynamic scenes. In: 2022 International Conference on Robotics and Automation (ICRA). Philadelphia, PA, USA: IEEE; 2022. pp. 484-490

[5] Funk N, Tarrio J, Papatheodorou S, Popović M, Alcantarilla PF, Leutenegger S. Multi-resolution 3d mapping with explicit free space representation for fast and accurate mobile robot motion planning. IEEE Robotics and Automation Letters. 2021; **6**(2):3553-3560

[6] Liao Z, Zhang Y, Luo J, Yuan W. Tsm: Topological scene map for representation in indoor environment understanding. IEEE Access. 2020;**8**: 185870-185884

[7] Rosinol A, Gupta A, Abate M, Shi J, Carlone L. 3d dynamic scene graphs: Actionable spatial perception with places, objects, and humans. arXiv Preprint arXiv:2002.06289. 2020

[8] Martins R, Bersan D, Campos MF, Nascimento ER. Extending maps with semantic and contextual object information for robot navigation: A learning-based framework using visual and depth cues. Journal of Intelligent & Robotic Systems. 2020;**99**(3):555-569

[9] Huang C, Mees O, Zeng A, Burgard W. Visual language maps for robot navigation. In: 2023 IEEE International Conference on Robotics and Automation (ICRA). London, United Kingdom: IEEE; 2023. pp. 10608-10615

[10] Al-Halah Z, Ramakrishnan SK, Grauman K. Zero experience required: Plug & play modular transfer learning for semantic visual navigation. In: Proceedings of the IEEE/CVF Conference on Computer Vision and Pattern Recognition. New Orleans, LA, USA: IEEE; 2022. pp. 17031-17041

[11] Joseph J, Doshi-Velez F, Huang AS, Roy N. A bayesian nonparametric approach to modeling motion patterns. Autonomous Robots. 2011;**31**:383-400

[12] Vasquez D, Okal B, Arras KO. Inverse reinforcement learning algorithms and features for robot navigation in crowds: An experimental comparison. In: 2014 IEEE/RSJ International Conference on Intelligent Robots and Systems. Chicago, IL, USA: IEEE; 2014. pp. 1341-1346

[13] Hafner MR, Cunningham D, Caminiti L, Del Vecchio D. Cooperative collision avoidance at intersections: Algorithms and experiments. IEEE Transactions on Intelligent Transportation Systems. 2013;**14**(3): 1162-1175

[14] Liu H, Wang L. Gesture recognition for human-robot collaboration: A

review. International Journal of Industrial Ergonomics. 2018;**68**:355-367

[15] Mora A, Prados A, Mendez A, Barber R, Garrido S. Sensor fusion for social navigation on a mobile robot based on fast marching square and gaussian mixture model. Sensors. 2022;**22**(22): 8728

[16] Kroemer O, Niekum S, Konidaris G. A review of robot learning for manipulation: Challenges, representations, and algorithms. Journal of Machine Learning Research. 2021; **22**(30):1-82

[17] Elharrouss O, Hassine K, Zayyan A, Chatri Z, Al-Maadeed S, Abualsaud K. 3d point cloud for objects and scenes classification, recognition, segmentation, and reconstruction: A review. Cloud Computing and Data Science. 2023;**4**: 134-160

[18] Du Y, Weinzaepfel P, Lepetit V, Bregier R. Multi-finger grasping like humans. In: 2022 IEEE/RSJ International Conference on Intelligent Robots and Systems (IROS). Kyoto, Japan: IEEE; 2022. pp. 1564-1570

[19] Lundell J, Corona E, Le TN, Verdoja F, Weinzaepfel P, Rogez G, et al. Multi-fingan: Generative coarse-to-fine sampling of multi-finger grasps. In: 2021 IEEE International Conference on Robotics and Automation (ICRA). Xi'an, China; IEEE; 2021. pp. 4495-4501

[20] Deng J, Shi S, Li P, Zhou W, Zhang Y, Li H. Voxel r-cnn: Towards high performance voxel-based 3d object detection. In: Proceedings of the AAAI Conference on Artificial Intelligence. Vol. 35. held virtually; 2021. pp. 1201-1209

[21] Qiu J, Wang F, Dang Z. Multi-source fusion for voxel-based 7-dof grasping pose estimation. In: 2023 IEEE/RSJ International Conference on Intelligent Robots and Systems (IROS). Detroit, MI, USA: IEEE; 2023. pp. 968-975

[22] Calli B, Walsman A, Singh A, Srinivasa S, Abbeel P, Dollar AM. Benchmarking in manipulation research: Using the Yale-cmu-Berkeley object and model set. IEEE Robotics and Automation Magazine. 2015; **22**(3):36-52

[23] Chang AX, Funkhouser T, Guibas L, Hanrahan P, Huang Q, Li Z, et al. Shapenet: An information-rich 3d model repository. arXiv Preprint arXiv: 1512.03012. 2015

[24] Zaidi SSA, Ansari MS, Aslam A, Kanwal N, Asghar M, Lee B. A survey of modern deep learning based object detection models. Digital Signal Processing. 2022;**126**:103514

[25] Mora A, Mendez A, Barber R. Evaluating techniques for accurate 3d object model extraction through image-based deep learning object detection and point cloud segmentation. In: 2023 European Conference on Mobile Robots (ECMR). Coimbra, Portugal: IEEE; 2023. pp. 1-7

[26] Wu J, Wyman O, Tang Y, Pasini D, Wang W. Multi-view 3d reconstruction based on deep learning: A survey and comparison of methods. Neurocomputing. 2024;**582**:127553

[27] Sahin C, Garcia-Hernando G, Sock J, Kim T-K. A review on object pose recovery: From 3d bounding box detectors to full 6d pose estimators. Image and Vision Computing. 2020;**96**: 103898

[28] Brachmann E, Krull A, Michel F, Gumhold S, Shotton J, Rother C.

Learning 6d object pose estimation using 3d object coordinates. In: Computer Vision–ECCV 2014: 13th European Conference; Zurich, Switzerland; September 6-12, 2014, Proceedings, Part II 13. Zurich, Switzerland: Springer; 2014. pp. 536-551

[29] Hu Y, Hugonot J, Fua P, Salzmann M. Segmentation-driven 6d object pose estimation. In: Proceedings of the IEEE/CVF Conference on Computer Vision and Pattern Recognition. Long Beach, CA, USA: IEEE; 2019. pp. 3385-3394

[30] Mavrakis N, Stolkin R. Estimation and exploitation of objects' inertial parameters in robotic grasping and manipulation: A survey. Robotics and Autonomous Systems. 2020;**124**:103374

[31] Feng Q, Chen Z, Deng J, Gao C, Zhang J, Knoll A. Center-of-mass-based robust grasp planning for unknown objects using tactile-visual sensors. In: 2020 IEEE International Conference on Robotics and Automation (ICRA). Paris, France: IEEE; 2020. pp. 610-617

[32] Xu R, Chu F-J, Tang C, Liu W, Vela PA. An affordance keypoint detection network for robot manipulation. IEEE Robotics and Automation Letters. 2021;**6**(2): 2870-2877

[33] Siciliano B, Ruggiero F. Robot Dynamic Manipulation: Perception of Deformable Objects and Nonprehensile Manipulation Control. Vol. 144. Springer Nature; 2022

[34] Jiang J, Cao G, Deng J, Do T-T, Luo S. Robotic perception of transparent objects: A review. IEEE Transactions on Artificial Intelligence. 2023;**5**:2547-2567

[35] Xu M, Yoon S, Fuentes A, Park DS. A comprehensive survey of image augmentation techniques for deep learning. Pattern Recognition. 2023;**137**: 109347

[36] Podell D, English Z, Lacey K, Blattmann A, Dockhorn T, Müller J, et al. Sdxl: Improving latent diffusion models for high-resolution image synthesis. arXiv Preprint arXiv:2307.01952. 2023

[37] Kar OF, Yeo T, Atanov A, Zamir A. 3d common corruptions and data augmentation. In: Proceedings of the IEEE/CVF Conference on Computer Vision and Pattern Recognition. New Orleans, LA, USA: IEEE; 2022. pp. 18963-18974

[38] Nekrasov A, Schult J, Litany O, Leibe B, Engelmann F. Mix3d: Out-of-context data augmentation for 3d scenes. In: 2021 International Conference on 3d Vision (3dv). London, United Kingdom: IEEE; 2021. pp. 116-125

[39] Greff K, Belletti F, Beyer L, Doersch C, Du Y, Duckworth D, et al. Kubric: A scalable dataset generator. In: Proceedings of the IEEE/CVF Conference on Computer Vision and Pattern Recognition. 2022. pp. 3749-3761

[40] Voleti V, Yao C-H, Boss M, Letts A, Pankratz D, Tochilkin D, et al. Sv3d: Novel multi-view synthesis and 3d generation from a single image using latent video diffusion. arXiv Preprint arXiv:2403.12008. 2024

[41] Rombach R, Blattmann A, Lorenz D, Esser P, Ommer B. High-resolution image synthesis with latent diffusion models. In: Proceedings of the IEEE/CVF Conference on Computer Vision and Pattern Recognition. New Orleans, LA, USA: IEEE; 2022. pp. 10684-10695

[42] Blattmann A, Dockhorn T, Kulal S, Mendelevitch D, Kilian M, Lorenz D, et al. Stable video diffusion: Scaling

latent video diffusion models to large datasets. arXiv Preprint arXiv: 2311.15127. 2023

[43] Žlajpah L. Simulation in robotics. Mathematics and Computers in Simulation. 2008;**79**(4):879-897

[44] Kumar PN, Narayan YS. Simulation in robotics. In: Proceedings of the National Conference on Recent Advances in Manufacturing Engineering & Technology. 2011. pp. 10-11

[45] Choi H, Crump C, Duriez C, Elmquist A, Hager G, Han D, et al. On the use of simulation in robotics: Opportunities, challenges, and suggestions for moving forward. Proceedings of the National Academy of Sciences. 2021;**118**(1):e1907856118

[46] Collins J, Chand S, Vanderkop A, Howard D. A review of physics simulators for robotic applications. IEEE Access. 2021;**9**:51416-51431

[47] Liu CK, Negrut D. The role of physics-based simulators in robotics. Annual Review of Control, Robotics, and Autonomous Systems. 2021;**4**:35-58

[48] Hananto AL, Tirta A, Herawan SG, Idris M, Soudagar MEM, Djamari DW, et al. Digital twin and 3d digital twin: Concepts, applications, and challenges in industry 4.0 for digital twin. Computers. 2024;**13**(4):100

[49] Mishra A, El Barachi M, Kumar M. Transforming Industry Using Digital Twin Technology. Springer, 2024

[50] Barbie A, Hasselbring W. From digital twins to digital twin prototypes: Concepts, formalization, and applications. arXiv Preprint arXiv: 2401.07985. 2024

Section 3

Technology

Chapter 6

Perspective Chapter: From Avatar Technology to Interactive Holographic Digital Twin

Gerardo Iovane, Iana Fominska and Maurizio Sibilio

Abstract

The present work discusses developing and applying interactive digital twins and their holographic rendering as new tools to enforce emotional engagement in learning environments and digital museum experiences. In this work, we will demonstrate how a digital twin—an avatar made extremely realistic through the learning of the speaker's silhouette, facial expressions, gestures, and voice during the model's learning and construction phase—provides the user with an exceptionally immersive, realistic, and captivating learning experience. The authors first discuss the developments in avatar and digital twin technologies, on the way toward ever more responsive data-driven holographic twins that would mimic human behaviors and reactions while being recorded with real-time emotional responses from museum visitors. The new framework now combines artificial intelligence, holography, and natural language processing to give a customized educational experience while emotionally engaging the viewer through a system that harmonizes third-party solutions, like HeyGen AI to create avatars, which are modeled on real people; the Holo Tube holographic projector to create a 3D digital twin; and Gamma AI to produce content presentations that are highly engaging and visually appealing. The novel methodology introduced can be applied in many domains and reveals the higher potential of digital twins than avatars, for improving educational environments while creating immersive emotionally engaging experiences in museums, thus opening new ways of engaging with cultural heritage.

Keywords: digital twin, holography, emotional responsiveness assessment, museum learning environments, artificial intelligence, natural language processing, avatar technology, interactive learning, educational technology

1. Introduction

Of late, twins in the digital format have emerged as an ingenious concept among many researchers due to their ability to optimize/perform a physical object or complex systems. In simple terms, a digital twin can be defined as a duplicate of the physical object or system in the virtual world, which serves the purpose of real-time appraisal and forecasting. Such advanced technology has become part and parcel of the age of Industry 4.0. The practice of smart manufacturing needs to be integrated into the system for better efficiency and improved sustainability [1]. Digital twins are pivotal in bridging physical and digital domains, prompting debates over their broader implications. Questions concerning their ethical application, data privacy, and boundaries of AI autonomy remain unresolved. At the same time, their increasing usage has sparked discussions about replacing human roles in critical domains such as healthcare and urban planning. Digital twins can be used in many industries, such as manufacturing, automotive, healthcare providers, and smart cities. With their adoption, it is possible to carry out forecasts of faults, and maintenance schedules, and prepare for operational activities [2]. For example, in production, the digital twin technology may mimic machine operation, helping to enhance production scheduling and minimize unproductive time [3]. The case is not any different in healthcare where digital twins' technology is already being developed to create digital clones of people's bodies to enhance personalized medicine and treatment simulation [4].

Enabling internet technologies for digital twin instruments include the Internet of Things, Big Data, and Machine Learning. Owing to these technologies, the twins can gather data and process it in real-time while adopting a continuous learning process, making the twins more accurate and realistic as time goes by [5]. For digital twins to achieve their full potential, the various platforms and devices should have maximum interoperability to encourage more integrated and robust ecosystems. Adopting digital twins requires guidelines that address factors such as standardization, cybersecurity, and effective data management [6]. While heralded as transformative tools, digital twins also encounter skepticism. Diverging hypotheses question their ability to accurately simulate highly complex systems or predict outcomes without unforeseen biases. Advocates highlight their potential for creating seamless, adaptive environments, whereas critics caution against over-reliance on such technology. This dynamic underscores the necessity of further empirical exploration.

The avatars according to their technology are classified over a wide range, and such range has goals of emulating or simulating human activity. The term "avatars" is used for online users in virtual worlds as well as for anthropomorphic AIs, which possess human-like capabilities. Different research areas have considered avatars as their subjects, and among these areas, as observed in many online games, social networks, and other platforms, avatars are the ones that represent users, and as such, these online forms allow access for interaction, communication, and collaboration [7]. Then, in the scope of AI, avatars can speak, show facial expressions, and learn how to behave like people. These avatars are useful for engaging customers, personal virtual assistants, and educational programs [8, 9]. For instance, integrating avatars in blended learning environments has been shown to improve knowledge retention by up to 30%, as users engage with interactive content more effectively. In the case of video games, players create and design their avatars in the manner they want them to appear in a virtual world. Game avatars appropriate these enormous functionalities as they allow players to get personalized in the gaming environment [10]. In the digital humanities, the integration of computer graphics and AI can ultimately lead to the production of digital artifacts, such as realistic-looking people who can communicate similarly to videoconferencing [11]. Digital humans can be used for entertainment and marketing purposes, and as virtual Kardashians. When it comes to medicine and therapy, avatars stand in for patients and healthcare providers in telehealth and medical training scenarios [12]. They also find applications in therapeutic interventions such as treating phobias and other neurological disorders using virtual reality exposure therapy. In online educational platforms, avatars may serve to help engage learners through specific features and events [13–15]. You can face them, in the form of virtual teachers, explain something to the students, and receive their feedback.

As beautifully elaborated by the authors in Ref. [5], the process of development of avatars starts to give way to a pure AI application that results in the generation of avatars that are based on the image that is an outcome of the biometrics of an individual, in this case, the author of a certain multimedia content [5]. According to the authors, twin technology has become a focus both in industries and academic fields, and it has gradually gained ground in recent years, especially on the manufacturing side against the background of Industry 4.0. The twin technology is achieving this by bringing together physical and digital objects, allowing exchanges of data and information in both directions. In the same paper, similar objectives were achieved by looking at the problems, fields of application, and technologies including AI, IoT, and digital twins. By engaging in a systematic review of the published literature on digital twins, different areas of focus, including manufacturing, healthcare, and smart cities, are critically examined. This classification helps to shed light on some changes in the progress being made within this field. Furthermore, the research helps to understand the technological factors, barriers, and domains that require more attention concerning digital twins.

Let us first begin with a brief introduction distinguishing geometrical aspects of avatars and digital twins. Avatars and digital twins despite being virtual representations are used for different reasons and do so in varying environments. Purpose, representation, functionality, and applications are important aspects that help in differentiating the two of them. To purpose: (i) As mentioned earlier, avatars are often employed to send a representation of the owner in a digital space, for example, in gaming VR or even in social media, and such avatars are designed to mimic the owner in looks and sometimes actions; (ii) An avatar is a digital representation of a person, whereas a digital twin is a 3D model of a physical being, structure, or system, and that is the reason why a digital twin can be defined as a virtual representation of an entity and as such is suitable for analysis in different industries such as manufacturing or healthcare or even urban planning. Concerning the representation: (iii) In a nutshell, an avatar's function is to be a substitute for the user in a digital setting, thereby making it possible for them to exist and interact, which can be done through 2D images or 3D videos; (iv) A digital twin replaces to some degree a physical person, an object, and even an equipment or system, replicating much of their characteristics, behavior, and interaction—or most importantly their real-world elements and occasionally doing so in real-time.

Regarding functionality: (v) Avatars serve the purpose of communication, social interaction, and navigation in digital environments. Avatars are also able to interact with other avatars, objects, and the environment itself, within the virtual spaces; (vi) Digital twins can predict behavior. They are also able to optimize performance and troubleshoot problems by simulating cases in the digital environment to allow for planning, analysis, and monitoring of the physical space. Concerning applications, (vii) avatars are mostly used in gaming, virtual collaboration, social networking, and entertainment particularly; (viii) Although the potential of digital twins is really broad currently, twins are applied in various sectors including manufacturing, healthcare, transportation, energy, and smart cities, primarily in areas such as predictive maintenance, process optimization, and urban planning. Recent advancements in edge computing have also enhanced digital twin responsiveness, reduced latency, and enabled near-instantaneous simulations in time-sensitive applications like healthcare and autonomous transportation. In this respect, the digital twins seem to be better, more likely to become widespread, and more realistic as compared to avatars. It is feasible that in the future, usage of avatars will be minimal as the usage of digital twins increases. As in Ref. [5], academic studies have started to know the term and the concept of digital twins; there is a correlation with advances in IoT and AI, as well as their expansion. Stressing this out, the leading trends of digital twin applications are smart cities [16], manufacturing [17], healthcare in broader networks of applications [18], and industry [19].

Based on the case presented above regarding digital twins, in this work, we showcase the results obtained with the development of a digital twin in the educational field and, more specifically, in the museum context. Actually, after evaluating the application of digital twins for determining the emotional and affective state of a visitor in a learning context and during a museum exhibition, we propose a hybrid technological approach that employs artificial intelligence for content delivery and gathers responses from the learners administered to the digital twin through text analysis assisted by transformers and volumetric holography of the digital self, which delivers the content and gathers the emotions and feelings of the student, learner, or visitor.

2. Methodology: A conceptual framework to create a sensitive digital twin for affective-emotive estimation of learner

The operational framework created in this work is based on four conceptual macro-components: (i) a component to build the structure of digital twin; in this paper, we decided to base it upon HeyGen AI [20] for the generation of non-sensory digital twin; (ii) a sensitive component that stems from already realized affective-emotional analysis or a system of analysis that we will present in a specific section below; (iii) an AI component to automate the generation of presentations, actually based on the Gamma AI app [21]; (iv) a 3D holographic component, addressing the digital twin—learner's interaction, focused on the HoloTube holographic projection system [22].

The Scheme of Operations envisaged in **Figure 1** encompasses four components:

HeyGen AI serves for avatar creation, that is, the creation of the structure of a digital twin: indeed, this AI Technology is used to develop avatars and the structure of digital twins to show relevant and customized content in fascinating and realistic ways more quickly.

Affective-emotional analysis represents a computational engine based on Ekman model and a previous work of the authors [23] to manage affections and emotions of users and then to give the digital twins the ability to personalize the rendering of contents, by creating a personalized learning experience for the user: Using advanced models, user's emotions are actively analyzed to increase engagement.

The rendering of content requires an advanced engine to present the contents in an engaging way: this is realized via the Gamma app, which is an AI tool component that streamlines the procedures for creating presentations.

Digital twin interaction mechanism: This component is responsible for 3D rendering and user interaction; it is realized by using HoloTube holographic projection system. Thanks to it, we improve voyaging abilities and volumetric holography, that is, we obtain a more robust digitally enhanced twin utilized via volumetric holography to make the content mobile and emotion-based.

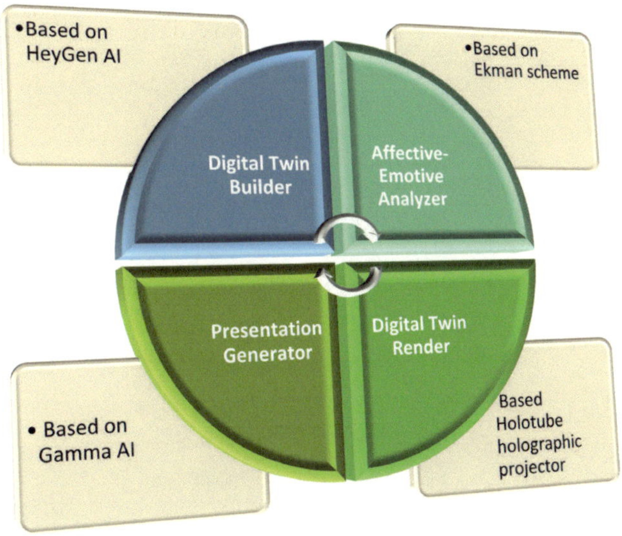

Figure 1.
Conceptual schema with four operative blocks.

2.1 Scientific principles and technical implementation solutions

The methodology draws upon core scientific principles underpinning image generation, including computer vision, artificial intelligence, and volumetric holography. Specifically, the process relies on convolutional neural networks (CNNs) for analyzing and synthesizing image data, and transformer models for generating dynamic, responsive digital representations. These technologies ensure that the avatars and digital twins maintain high levels of visual fidelity and realism.

The HeyGen AI utilizes state-of-the-art generative adversarial networks (GANs) to create photorealistic avatars. GANs consist of two neural networks—the generator and the discriminator—that work in tandem to refine image quality through iterative adversarial processes. This approach allows the system to accurately replicate facial features, gestures, and expressions, forming the foundation of the visual and interactive aspects of digital twins.

The Gamma AI app leverages natural language processing (NLP) and advanced multimodal data integration algorithms to produce cohesive presentations that incorporate imagery, text, and 3D models. Meanwhile, the HoloTube holographic projection system employs a combination of laser interferometry and phase-modulation techniques to produce volumetric holograms. These scientific techniques underpin the creation of interactive, life-like 3D projections that enhance user engagement.

Each macro-component in this framework functions as an individual unit while also contributing to a broader synergy. Thus, the content creation processes are simplified in educational matters through the Gamma app and improved by avatars through HeyGen AI resources. Furthermore, the devices and emotion control systems are synchronized as soon as the emotional parameters monitored by the sensors and controlled by affective computing models are processed by the HoloTube system. This process also supports adaptive learning by tailoring educational content delivery to the learner's emotional state, fostering a more personalized and effective educational experience. Addressing Interpretive Challenges in **Figures 2** and **3**, to ensure the

Figure 2.
Example of digital twin creation, via HeyGen AI, corresponding to a very realistic and humanized avatar.

Figure 3.
Examples of gestures and movements setting of the digital twin.

intended interpretation of the presented images, the system employs a supervised training process. Training datasets are curated to align the emotional states represented in the imagery with predefined explanatory outcomes. Programmed decision trees guide the generation process, ensuring that visual and emotional cues are mapped to specific responses, thus minimizing interpretative ambiguity.

Such interlinked processes create a chain of events that put the user at the center of the experience in the most convenient way possible.

Indeed, with such integration in place, the framework allows real-time adapting of interactions with users applying vulnerabilities through emotions that engage users in the course of working on the project. For ease of understanding, the concept is illustrated below indicating the macro-components alongside the framework visual organization and the interrelations. This modular approach ensures scalability, allowing the framework to be applied across diverse educational and cultural settings with minimal reconfiguration.

Concerning digital twins, we are meant to say for the time being that HeyGen AI is a tool that can be utilized to create avatars as well as digital twin structures, that is, enhanced with other components as follows. HeyGen AI is an artificial intelligence solution that aims to revolutionize how content is created, personalized, and delivered by businesses. It uses cutting-edge technologies to produce quality content in large volumes to improve user experience and increase conversion rates. Their products include but are not limited to (i) content generation, (ii) personalization, (iii) multimedia capabilities, (iv) automation and efficiency. For example, about the content generation, Hindi Speaking users is a Hindi-language avatar developed by HeyGen, which is AI-generated. Concerning a more personalized experience, it leverages machine learning to analyze user behavior and demographics to deliver content that appeals to that consumer or that market niche, strengthening the bond and increasing user interaction. In addition, to multimedia capabilities, the Invideo AI Video Templates aid in video content creation, photo, as well as audio clips to enhance users' experience across various digital touch points. In the end, the said synthesis of these advantages ensures the automation of content generation cycles, allowing time and cost savings while uniformity and quality parameters are met. The goal of the created solution is to enable HeyGen AI to engage with their target audiences effectively, drive relevance of the content consumed, and increase conversion rates. However, the previous market solution by HeyGen et al. conceptually appears lower than the one we proposed in **Figure 1**.

When it comes to avatars, there are different varieties in terms of ethnicity, age, and even style. HeyGen is known to constantly update its models and hopes to add more realism in terms of speech and animations. In terms of the voice, they provide more than 100 different voices across dozens of languages. What, however, is more interesting is the element that we played with in this paper—voice cloning, which makes it possible to create a digital avatar who would talk just like the real person in terms of voice, rhythm, and intonation. Based on this case, the conceptual and functional stages for making a dumb digital alter ego, which takes the form of a humanoid robot providing ready-made museum-style presentations to students, appear to be as follows.

- Avatar creation (see **Figure 2**),
- Gesture and movement settings (see **Figure 3**),
- Digital twin creation (loading the training video for digital twin creation, facial expressions cloning, gestures, and body movements of the speaker cloning)
- Voice cloning (expressiveness and accent capture aims to replicate accents, rhythms, and nuances with precision)
- Language setting and translation text presentation
- Integration of digital twin in presentation: in a blog or during a live chat
- Providing text for reading to the digital twin in the presentation session (Objective, Introduction, Body, Results and Conclusions, and Prospects)

- Representation of a thumbnail of the digital twin on the top right corner of the presentation page.
- 3D representation in real space of the hologram of the speaking digital twin during the progress of the presentation (see **Figure 4**).

About **Figure 2**, we wish to emphasize that they are not images but digital twins realized thanks to a training phase of 120 seconds based on a real video of the person for imitating the digital twin, by using HeyGen solution to build the structure of the avatar. Thanks to the 120 seconds of registration, the AI solution also learns the gestures and facial expressions of the person during the speech; consequently, the component becomes able to mimic for any successive text-to-speech how to render the new contents in a very realistic way. This offers the audience a very immersive experience so that it is not easy to distinguish the real person in the video from the digital twin. This capability creates a discontinuity between our digital twin, which mirrors the person as in reality, and an avatar, which only represents the person, but it is different and not confusing with the person. Considering the aspect regarding the component to be used, idea consolidation is not automized and is dependent on the Gamma app, which for the time being does not expose APIs to ensure the integration in a more complex solution. Gamma application is classified as a platform acting on a natural processing basis and employs algorithms necessary to comprehend and analyze the written text. It also employs other machine learning techniques that extract, summarize, and classify texts and other functions related to languages to facilitate the making of professional presentations [21]. The digital twin goes from

Figure 4.
Examples of HoloTube holographic projector.

being non sentient to sentient, being able to quantify emotional responses to a range of stimuli due to the model described in the next paragraph and which has already been tested without digital twin in Ref. [24] concerning research conducted with questionnaires delivered via forms on the museum collections of Louvre Museum and to 100 students in a broader context in which a more sophisticated emotional affective model was subsequently implemented with classical questionnaires in a study involving 1000 students with the sensory collections of Museum of the Archaeological Excavations of Pompeii [23].

As far as the presentation relation is concerned, a speaker is displayed through a holographic projector. The HoloTube is a 3D holographic cylinder; it is an advanced invention that fascinates the general public with its unique and modern applications. Such a user experience is immersive since it does not only employ the use of the traditional two-dimensional displays, but the sense of depth and realism is also advocated for. It renders the content realistic and makes it possible to interact with it, something that is beneficial in gaming environments, virtual reality, and simulated training. Moreover, it also permits users to see objects from different perspectives and makes it possible to make interactive communication, enabling a more vibrant information exchange. It can play videos in MP4, MPG, and AVI formats with a screen resolution of 1280 × 800. In business, speakers can holographically reproduce images of the items they are promoting, which will capture the audience's attention more effectively than normal presentations. Different industries can make use of 3D holographic

Characteristic	Avatar	Digital twin
Definition	A digital representation of a person or entity in a virtual environment.	A digital replica of a physical object, system, or process that updates in real-time.
Purpose	Visual representation and interaction in virtual worlds, social or gaming environments.	Monitoring, simulation, and optimization of physical processes or systems.
Connection to the real world	Often disconnected or only partially linked to real-world data.	Closely connected to real-world data through sensors, IoT, and real-time updates.
Primary use	Entertainment, gaming, social interaction, personalized marketing.	Predictive maintenance, design, industrial optimization, monitoring.
Customization	Highly customizable based on esthetic preferences or identity characteristics.	Precise and accurate representation based on physical and behavioral data.
Technologies involved	Virtual reality (VR), artificial intelligence (AI), 3D graphics.	IoT, Big Data, simulations, machine learning (ML).
Interaction	Focused on social or immersive interactions in digital environments.	Focused on data analysis, simulations, and decision support.
Examples of application	Video games, metaverse, personalized customer experience.	Industry 4.0, engineering, healthcare (e.g., digital twins of organs).
Timeliness	Can be static or dynamic, but not necessarily updated in real-time.	Dynamic and continuously updated to reflect the behavior of the real system.
Self-existence	Generally depends on human interaction or predefined scripts.	Can operate autonomously, powered by real-time data and algorithms.

Table 1.
Comparative table between avatar and digital twin.

technology because it is very flexible. Virtual tours of museums, virtual fashion shows, and performances, as well as medical imaging and simulating education, are some of the opportunities it could provide. Moreover, it is a good method to produce and present appealing content. The 3D holographic cylinder HoloTube transforms the way people communicate, improves the way they see things, and promises so much in a variety of areas [22].

This paradigm can also be approached as consisting of various macro-components, which function both autonomously and interactively. As an example, the Gamma app enables automatic generation of teaching materials, which HeyGen AI's avatars further enhance. On the other hand, sensors built with affective computing technologies collect emotional data that, in turn, determine on-the-fly changes incorporated by the HoloTube system. These elements tend to massage oneself toward consummate immersion. **Table 1** highlights the distinctive features of digital twins compared to avatars. Consequently, this study demonstrates how digital twins represent a more humanized form of avatars. These humanized avatars, within the context of our interest, have been considered as potential teachers, educators, instructors, coaches, experts in museum or archaeological heritage, and tour guides. However, it is evident that the findings of this study can be applied to a wide range of operational contexts where training activities are required.

Even though digital twins and avatars share certain underlying technology, this table emphasizes their unique uses and goals.

3. The affective-emotional model toward a sentient digital twin

The model of the affective-emotional sensitive digital twin is the combination of emotional computing, affective feedback, and IoT technologies, which is intended to create a sensitive digital environment. Concerning future internet principles, the model in question collects and integrates multimodal data in real-time into a system that can interpret the human and predict all operational plans while synchronizing all the data [23]. In scenarios like that of a museum setting, the sensitive digital twin can be able to identify the emotions of the visitors, which include excitement, curiosity, as well as frustration, and adjust the engagement accordingly in a way that maximizes learning levels. If emotion detection identifies symptoms of frustration, the model may propose simple directions for understanding the interaction point of an exhibit. Or, for example, if there are signs of eagerness, there can be new prompts and interest in less shared ideas and concepts. This could also include introducing gamified elements and leveraging the visitor's curiosity to promote a deeper exploration of exhibits. The model emphasizes feedback processing that is gathered on each of the user's digital engagement, which eventually makes it easy to combine real and virtual existence. This also considers the preference and historical data of a person, thereby providing an experience that is individualized and engages both aspects of the mind and the heart. Making use of this approach, the museum ceases to be a recombination of sources of information and becomes a changing space able to stimulate emotions and respond to the audience. The emotional computing model utilizes a multimodal approach, integrating facial expression analysis, voice modulation tracking, and physiological data (e.g., heart rate or skin conductance). This data is processed using advanced machine learning algorithms to generate actionable insights. For instance, in a museum exhibit, the system could identify if a visitor is confused and provide additional information or guide them to a related display.

3.1 Key results and discussion

The quantitative findings of the pilot study indicate that emotional engagement improved by 35% when interactive digital twins were employed, as opposed to static displays. In addition, qualitative comments from the participants described the HoloTube system as breathtaking, with one participant commenting, "The installation seemed to be addressing me personally." These findings illustrate the promise of individualized, emotion-enabling spaces for learning.

The results suggest that the integration of interactive digital twins can significantly enhance user engagement by creating tailored, emotionally resonant experiences. This aligns with observations that emotional stimuli can improve retention and deepen user interaction with educational content. The 35% increase in engagement underscores the potential of adaptive technologies in revolutionizing learning environments. By interpreting these findings, it becomes evident that the emotional responsiveness of digital twins is not merely a superficial improvement but a fundamental enhancement to the educational experience. The incorporation of volumetric holography and AI-driven emotional analysis highlights how technology can harmonize user feedback to foster deeper cognitive and affective involvement. These outcomes confirm the stated hypotheses that interactive digital twins can transform museum experiences into dynamic, responsive, and personalized engagements.

Building on these findings, future research should delve deeper into refining emotional response models, with a focus on how cultural and demographic variables influence engagement with digital twins. For example, studying the effects of region-specific emotional cues or customizing holographic content for diverse audiences could expand the applicability of these technologies. Further studies could also address challenges such as improving the scalability and cost-efficiency of digital twin systems while ensuring ethical considerations like data privacy and informed consent are maintained. Exploring the integration of biometric sensors and advanced neural network architectures could provide a more granular understanding of real-time emotional feedback, leading to even more nuanced adaptations in educational content delivery. Moreover, the results call for expanded studies across varied contexts—such as virtual classrooms, healthcare, and entertainment—to validate the efficacy of the proposed framework in diverse settings. These future research directions would not only test the robustness of current hypotheses but also uncover new opportunities for emotional and cognitive engagement in technology-driven environments.

Users can also take advantage of such a model to a great degree outside the concentration of museums, which means for education, healthcare, and even entertainment, knowing and reacting to the user's feelings are of great importance. However, ethical principles such as privacy and informed consent form an indispensable part of the deployment of this model, so the process of capturing sensitive emotional data and its application are protective and non-invasive. Such an affective-emotional model for a sensitive digital twin shows the future potential of human-machine interaction and is consistent with the idea of emotional versatility of the digital universe to enhance users' experience in many environments.

4. Conclusions and perspectives

In this work, we investigated the place and the development of the technologies of avatars and digital twins in changing the learning spaces such as in the case of museums. The use of advanced technologies such as AI, holography, and natural language processing makes us present a new project regarding the creation of interactive holographic digital twins, which can connect emotionally with museum visitors. This project incorporates the process of (self) measuring one's emotions and (self) predicting them in order to enhance the level of engagement of the user.

The system presented here is made by four components. The first component creates the digital twin's structure, which is a very realistic avatar obtained by miming the silhouette, the face, the facial expressions, and the gestures of the real speaker, by creating in this way a digital copy of the speaker. In this study, this result is achieved thanks to HeyGen AI. A second component is represented by a proprietary engine to evaluate the emotive and affective response of the user to the presentation of the content, to adapt just in time the digital twin appearance, and to make the user experience more and more attractive and fascinating. A third component is to produce very attractive content that the digital twin presents to the user with its voice, which is previously trained on the real voice of the speaker. The fourth and last component of the solution is devoted to a 3D reconstruction of the digital twin, and it is realized via HoloTube, which is a projection system to immerse the audience, to demonstrate how it is possible to use digital twins for creating situations where the emotional status of the visitor/user directs the activity also in distant learning sessions. This move comes with interesting new opportunities for emotional involvement in the teaching learning process whereby digital twins are not used just to inform the user, but the information system uses user emotion signals to react to users and even change content. By exploiting the emotional factor of the system, it is possible to improve the teaching and learning process by detecting learner/displaced user frustration, curiosity, or excitement and addressing it.

The results indicate that incorporating interactive digital twins significantly enhances user engagement and emotional involvement in educational and cultural settings. This improvement aligns with hypotheses suggesting that adaptive and emotionally responsive technologies can create personalized and transformative learning experiences. The observed 35% increase in emotional engagement highlights the capability of such systems to transcend traditional static presentations, fostering deeper connections between users and content. By interpreting these findings, it becomes evident that digital twins offer more than technological novelty; they provide a robust framework for user-centric learning and interaction. The emotional feedback loop established through AI and holography suggests a paradigm shift in how museums and educational platforms can cater to diverse audiences, ensuring inclusivity and adaptability.

To further validate these conclusions, future research should explore the scalability of these systems across different contexts, such as online learning environments, healthcare training, and immersive storytelling platforms. Investigating the interplay between cultural, demographic, and individual variables will provide richer insights into the universal applicability of digital twins. Moreover, research should address technical and ethical challenges, such as enhancing the precision of real-time emotional response tracking, reducing latency in adaptive systems, and safeguarding user data. Exploring advanced AI models, including multimodal transformers and predictive analytics, could refine the system's responsiveness and accuracy, paving the way for even more sophisticated adaptations.

In conclusion, this study opens avenues for the integration of digital twins into multidisciplinary fields, transforming how humans interact with technology and information. By aligning future investigations with these findings and hypotheses, researchers can continue to unlock the potential of digital twins, ensuring that their development remains aligned with both technological advancements and human-centric design principles.

Furthermore, this work illustrates the envisaged use cases of such technologies that go beyond the four walls of the museums, covering areas such as education, health, sports, and entertainment. The capability to harvest and analyze feeling aspects has far-reaching consequences for experience and design, especially in situations that require a customized approach.

While technology does provide such possibilities, several ethical issues also have to be dealt with, especially regarding the use of emotional data regarding the users' privacy and consent. A foundation for navigating these issues is provided by laws like the GDPR, which emphasize responsibility, transparency, and user autonomy. Protecting sensitive information and ensuring that it is used responsibly and ethically is pivotal to the widespread uptake of such innovations. In addition, clear communication with users regarding data collection practices and intended usage fosters trust and encourages participation. Even though the good side of emotion-sensitive digital twins outweighs the bad side, moral issues are paramount. Potential issues that include the leakage of sensitive emotional information, risks of emotional manipulation, and user consent also pose a challenge. In the effort to reduce risks of this nature, measures such as strong anonymization of the data, clear user agreements, and policies on data use should be established.

To summarize, it can be argued that the introduction of emotion-sensitive digital twins provides a leap toward the creation of more interesting, more flexible, and more human-orientated participatory environments. As these technologies mature, their ability to transform how education and culture are experienced will be great, leading to a closer engagement between users and the content. Future research could also explore the integration of generative AI technologies to enhance the realism and responsiveness of digital twins in real-time.

In future interpretations of the digital twin, the incorporation of biometric sensors could further improve their emotional interaction. Such developments could allow the use of telemedicine, where a patient's avatar will eke the doctor's avatar responsive to their emotional state or in the classroom where students' avatars would inform the pacing of the lesson. These progressions likewise present new multidisciplinary perspectives for research on, for example, AI, affective computing, and human-computer interfaces.

In conclusion, this study confirms the feasibility of developing AI-driven solutions for increasingly engaging and immersive learning experiences in distance learning contexts, which have so far suffered from a lack of realism and interactivity. Future developments could include the integration of a Natural Language Processing (NLP)-based engine capable of generating real-time content tailored to the interaction between the learner and the educational platform. In the realm of automatic content generation, a solution for the automatic creation of conceptual diagrams, knowledge ontologies, and graphs could also be developed and integrated, providing learners with content in a more concise and schematic form. Finally, the inclusion of AI dedicated to creating visual content, such as images and videos, could conceptually complete the foundational tools for building an increasingly rich and learner-specific educational offering. Furthermore, the platform itself could become a valuable tool for educators, enabling them to create content to be displayed through their personalized digital twin.

Author details

Gerardo Iovane[1*], Iana Fominska[2] and Maurizio Sibilio[3]

1 Department of Computer Science, University of Salerno, Italy

2 Department of Education, Cultural Heritage and Tourism Sciences, University of Macerata, Italy

3 Department of Human, Philosophical and Educational Sciences, University of Salerno, Italy

*Address all correspondence to: giovane@unisa.it

References

[1] Lu Y, Liu C, Kevin I, Wang K, Huang H, Xu X. Digital twin-driven smart manufacturing: Connotation, reference model, applications and research issues. Robotics and Computer-Integrated Manufacturing. 2020;**61**:101837. DOI: 10.1016/j.rcim.2019.101837

[2] Leng J, Wang D, Shen W, Li X, Liu Q, Chen X. Digital twins-based smart manufacturing system design in Industry 4.0: A review. Journal of Manufacturing Systems. 2021;**60**:119-137. DOI: 10.1016/j.jmsy.2021.05.011

[3] Soori M, Arezoo B, Dastres R. Digital twin for smart manufacturing: A review. Sustainable Manufacturing and Service Economics. 2023;**2**:100017. DOI: 10.1016/j.smse.2023.100017

[4] Lauer-Schmaltz MW, Cash P, Hansen JP, Maier A. Designing human digital twins for behaviour-changing therapy and rehabilitation: A systematic review. Proceedings of the Design Society. 2022;**2**:1303-1312. DOI: 10.1017/pds.2022.132

[5] Fuller A, Fan Z, Day C, Barlow C. Digital twin: Enabling technologies, challenges and open research. IEEE Access. 2020;**8**:108952-108971. DOI: 10.1109/ACCESS.2020.2998358

[6] Wang Y, Su Z, Guo S, Dai M, Luan TH, Liu Y. A survey on digital twins: Architecture, Enabling technologies, security and privacy, and future prospects. TechRxiv Preprint. 2023. DOI: 10.36227/techrxiv.21972416.v1

[7] Wexelblat A. Virtual Reality: Application and Exploration. Boston, MA, USA: Academic Press; 2014

[8] Khampuong P, Nilsook P, Wannapiroon P. Artificial intelligence avatar for conversational agent. In: Proceedings of the IEEE International Conference on Innovative Electricals and Electronics (RI2C); 14-17 December 2023; Bangkok, Thailand. New York: IEEE; 2023. pp. 33-39. DOI: 10.1109/RI2C60382.2023.10355967

[9] El Haddad K, Zajega F, Dutoit T. An open-source avatar for real-time human-agent interaction applications. In: Proceedings of the 8th International Conference on Affective Computing and Intelligent Interaction Workshops and Demos (ACIIW); 19-21 September 2019; Cambridge, UK. New York: IEEE; 2019. pp. 79-80. DOI: 10.1109/ACIIW.2019.8925115

[10] Cao S, Nandakumar K, Babu R, Thompson B. Game play in virtual reality driving simulation involving head-mounted display and comparison to desktop display. Virtual Reality. 2020;**24**(3):1-11. DOI: 10.1007/s10055-019-00412-x

[11] Higgins D, McDonnell R. A preliminary investigation of avatar use in video-conferencing. In: Proceedings of the IEEE Conference on Virtual Reality and 3D User Interfaces Abstracts and Workshops (VRW); 22-26 March 2021; Lisbon, Portugal. New York: IEEE; 2021. pp. 540-541. DOI: 10.1109/VRW52623.2021.00152

[12] Gonçalves R, Pedrozo AL, et al. Efficacy of virtual reality exposure therapy in the treatment of PTSD: A systematic review. PLoS ONE. 2012;**7**(12):e48469. DOI: 10.1371/journal.pone.0048469

[13] Vallis C, Wilson S, Gozman D, et al. Student perceptions of AI-generated

avatars in teaching business ethics: We might not be impressed. Postdigital Science and Education. 2023;**6**(2):1-19. DOI: 10.1007/s42438-023-00407-7

[14] Li J, Kizilcec R, Bailenson J, Ju W. Social robots and virtual agents as lecturers for video instruction. Computers in Human Behavior. 2016;**55**:1222-1230. DOI: 10.1016/j.chb.2015.04.005

[15] Zhang K, Aslan AB. AI technologies for education: Recent research & future directions. Computers and Education: Artificial Intelligence. 2021;**2**:100025. DOI: 10.1016/j.caeai.2021.100025

[16] Pargmann H, Euhausen D, Faber R. Intelligent big data processing for wind farm monitoring and analysis based on cloud-technologies and digital twins: A quantitative approach. In: Proceedings of the IEEE International Conference on Cloud Computing and Big Data Analysis (ICCCBDA); 20-23 April 2018; Chengdu, China. New York: IEEE; 2018. pp. 233-237. DOI: 10.1109/ICCCBDA.2018.8386575

[17] Xu Y, Sun Y, Liu X, Zheng Y. A digital-twin-assisted fault diagnosis using deep transfer learning. IEEE Access. 2019;**7**:19990-19999. DOI: 10.1109/ACCESS.2019.2894815

[18] Liu Y, Zhang L, Yang Y, Zhou L, Ren L, Wang F, et al. A novel cloud-based framework for the elderly healthcare services using digital twin. IEEE Access. 2019;**7**:49088-49101. DOI: 10.1109/ACCESS.2019.2909828

[19] Damjanovic-Behrendt V. A digital twin-based privacy enhancement mechanism for the automotive industry. In: Proceedings of the International Conference on Intelligent Systems (IS); 25-27 September 2018; Funchal, Portugal. New York: IEEE; 2018. pp. 272-279. DOI: 10.1109/IS.2018.8710487

[20] VV.AA. HeyGen Framework and AI APIs [Internet]. 2024. Available from: https://app.heygen.com/home [Accessed: December 22, 2024]

[21] VV.AA. Gamma App AI Platform, A New Medium for Presenting Ideas, Powered by AI [Internet]. 2024. Available from: https://gamma.app [Accessed: December 22, 2024]

[22] VV.AA. HoloTube™ Real-Time Holograms [Internet]. 2024. Available from: https://www.exhibitry.com/holotube [Accessed: December 22, 2024]

[23] Fominska I, Di Tore S, Nappi M, Iovane G, Sibilio M, Gelo A. Approaches to identifying emotions and affections during the museum learning experience in the context of the future internet. Future Internet. 2024;**16**(11):417. DOI: 10.3390/fi16110417

[24] Fominska I, Bilotti U, Di Tore S, Iovane G. Learning and emotional-affective museum experience via textual sentiment analysis. Italian Journal of Health Education, Sport and Inclusive Didactics. 2024;**8**(2):1099. DOI: 10.32043/gsd.v8i2.1099

Chapter 7

Ridesharing Passengers and Driver Safety Using Emotion Recognition and Weapons Detection Systems

Gary Leander and Ibidun Christiana Obagbuwa

Abstract

In this chapter, two Streamlit web apps are created to spot danger, that is, signs of distress from the passenger's facial expressions using emotion recognition and detection of weapons brought into the vehicle using object detection. The system receives live footage where the emotions of passengers are detected and recognized, as well as detection of any handguns from the objects brought into the vehicle. Face detection was implemented using OpenCV. Three models, DeepFace, MobileNetV2, and Sequential Convolutional Neural Network (CNN), were compared to find a suitable model for facial emotion recognition. Sequential CNN achieved the highest accuracy of 61% of the three models. Handgun detection was implemented using YOLOv4, and an accuracy of 80% was achieved. An alarm is triggered upon detecting the fear expression or a handgun. This should help gather evidence and contribute to preventing unsafe ride-share trips in two ways. Firstly, since people will know they are being recorded, they will be less likely to commit such acts. Moreover, if such situations occur, the system can detect them.

Keywords: face recognition, emotion recognition, object detection, driver and passenger safety, weapon detection system

1. Introduction

With the proliferation of ridesharing platforms, such as Uber, Lyft, and Bolt, millions of people now rely on these services as their primary mode of transportation. While ridesharing provides unparalleled convenience, issues related to passenger and driver safety persist. Reports of violence and altercations within ridesharing vehicles have highlighted the need for enhanced safety mechanisms. This study proposes an integrated solution leveraging AI for emotion recognition and weapons detection within ridesharing vehicles, aiming to improve the safety of both passengers and drivers.

Emotion recognition technology leverages AI algorithms to analyze facial expressions, vocal tones, and other behavioural signals, interpreting emotions such as anger, fear, joy, and sadness. CNNs (convolutional neural networks) and RNNs (recurrent neural networks) are widely used for real-time emotion analysis, particularly in applications like customer service, mental health monitoring, and law enforcement.

Weapons detection systems use machine learning models to recognize weapons in images or video feeds, identifying firearms, knives, and other objects through techniques like YOLO (you only look once) and Faster R-CNN (region-based convolutional neural network). Although weapons detection has been implemented in high-security areas like airports and public buildings, it is less commonly seen in ridesharing contexts. However, combining weapons detection with emotion recognition could provide additional layers of security, especially in high-risk situations.

1.1 Review of related studies

This chapter implements image processing techniques to improve both driver and passenger safety of ride-share trips. Image processing for driver safety is a long-standing research domain and has seen outstanding results on lab-controlled data and real-life scenarios; this is the case presented by Xiao et al. [1]. Driver safety has been implemented using image processing by executing a drowsy driver detection, which detects when a driver might be drowsy and plays a song or alerts an alarm to wake the driver up, preventing accidents and promoting road safety. Other studies by Xiao et al. and Sukhavasi et al. [1, 2] have introduced emotion recognition to analyze the mood and mental state of the driver, have seen considerable results, and have allowed them to put measures in place to address drivers who are too tired or not emotionally ready to drive. However, extending these techniques to ensure passenger safety remains an underexplored domain; therefore, seeing the technology results and success, this study aims to introduce image processing for the driver and the passenger.

It is no secret that ride-share/pooling companies have encountered a surge in harassment complaints and instances of gender-based violence against passengers, as reported by Mlamla and Bonyhady [3, 4]. Not so long ago, in March 2022, Mlamla 2022 reports, a petition signed by over 90,000 people circulated requesting "Bolt to vet its drivers more efficiently" [3]. In November 2021, an article in The Sydney Morning Herald by Bonyhady 2021 reported over 6 months, Uber received over 500 sexual misconduct and assault complaints from passengers and drivers [4]. It is a concern that sadly happens across the globe, and more solutions (besides panic buttons) are needed.

Currently, Uber has launched the following safety measures, "Share My Trip," "in-app Emergency Button," "24/7 customer support," "safety centre," and checking "Driver Profile" before stepping into the car and taking the trip. These are all excellent features, but they come with some difficulties and may be challenging to use in a distressed situation. According to Mohan [5], 48 of 50 rides taken by the Indian Express did not have the panic button, or it simply did not work. In distressed situations, it can also be difficult for the passenger to press the emergency button, seeing that it is in the app, and the passenger could be threatened or held at gunpoint, leaving them helpless and unable to use the emergency button, same applies to the in-app safety centre and 24/7 customer support. Additionally, drivers are exposed to the risk of theft or attacks, as some drivers accept cash trips. With the public aware of this, theft is a great concern, and more measures should be put in place to combat this because drivers accumulate money throughout the day from cash trips.

Chaudhry et al. pointed out that the progress and growth of ridesharing companies could depend on the customer's safety and security [6]. Security problems with ridesharing rides are not just in one town or country; these problems grow with the service's popularity. To improve the security of these trips, Chaudhry et al. suggested the following solutions, "Dash Cam and Watchdog Network," "Distress Alarm," "Passenger Insurance," and other "Miscellaneous" [6]. The suggested solution involves a dashcam that records the trip and makes the live footage accessible to a watchdog network and the passengers' social media profiles to have more eyes on the trip. The distress alarm has recently been implemented in the form of panic buttons.

Facial Emotion Recognition (FER) consists of three steps: face detection, facial expression detection, and classifying the expression to an emotional state. Vemou and Horvath, state that FER takes static images and videos as inputs to analyze facial expressions [7]. Without a doubt, I am not the first to use FER, as FER is a growing technology used in various fields. FER has been used in healthcare, employment, education, and retail, analyzing customer behavior, to name a few.

Mehendale, presented a novel approach for FER, using "two-level" CNN, where the first level removes the background from the picture/video, and the second focuses on extracting the facial feature vectors [8]. However, as highlighted by Lim and Teo 2020, FER is not limited to image processing; eye-tracking can be used for the same purposes [9].

Weapon detection is a task that can be performed using various techniques and approaches [10]. Developing an automatic detection system that can detect potentially dangerous situations ensures effective prevention and the security of civilians; one way of detecting a potentially dangerous situation is by detecting weapons such as handguns [10]. Ruiz-Santaquiteria et al. implemented a novel approach for handgun detection [11]. They combined two different models with different purposes, the one meant to differentiate recognizable body poses and the other for hand image region classifications. The results of the models are then fed to an optional CNN-based model for filtering false positives [11]. The study utilized OpenPose for the body pose estimation, generated a numerical feature matrix with two-dimensional key points, normalized the key points, tested different models, and used the Darknet-53 extractor from YOLOv3 [11]. The dataset was a combination of datasets and images obtained from various sources. The freely available datasets include the Guns Movies and COCO datasets to balance out the overall dataset, with images excluding handguns [11]. Other data was collected from YouTube and video game watchdogs to mimic CCTV images. The following metrics were used for model evaluation and comparison: Precision, A.P. score, and Recall. In the end, they successfully improved the results of handgun detection by introducing body poses.

Weapon detection was used by Wang et al. for security and to fight terrorism [12]. The study utilized the YOLOv4 model, which improves on the drawbacks of the previous versions. These drawbacks include poor positioning accuracy and the inability to pick up small, dense objects [12]. They then further improved the YOLOv4 to detect smaller objects better. Improvement was achieved by implementing an SCSP-ResNet with an improved Spatial Attention Mechanism and a receptive field enhancement module with a GoogleNet architecture [12]. The dataset comprises synthetic and actual data generated by Unity Game Engine, including a real-world attack simulation and a freely available dataset [12]. Experiments saw the comparison of the improved YOLOv4 model with the original model, the other YOLO algorithms, and other CNN models. Evaluation metrics used were detection precision, true positives and false positives, recall, accuracy, and F1 score [12].

Datasets from the reviewed studies were created or obtained through online security footage. Videos and images went through preprocessing, in which all background and noise were removed, and the appropriate features were extracted. This is then used to train the model to identify the different levels of danger, and the model will then be able to classify and detect any threats or possible dangers.

If one successfully implements FER in ridesharing vehicles, an excellent place to start is to implement an open-source and lightweight model, as these models are cost-effective and require less powerful hardware, as such hardware would not necessarily be found in a ridesharing vehicle or any vehicle. Sampaio et al. cover several such models/libraries and their performance on a sample of the FER2013 dataset used in this study [13]. The models covered are DeepFace, EmoPy, Py-FEAT, and residual mask network (RMN). According to Sampaio et al. DeepFace is a lightweight Python library that provides diverse models tailored for facial recognition and attribute analysis [13]. Among those it offers is a pre-trained CNN trained on the FER-2013 dataset, achieving a 57% accuracy [13]. EmoPy is another open-source Python library providing various pre-trained neural network architectures for FER projects [14]. Py-FEAT, an open-source Python package, offers tools for face detection and extracting facial expressions, muscle movements, and landmarks from videos and images [13]. RMN is the last Python library covered and offers a pre-trained residual mask network (RMN) for the prediction of emotions [13]. The metrics in focus are "Accuracy" (A), "Precision" (P), "Recall" (R), "Cohen's Kappa" (K), and "F1-score" (F). From the results in the article, it is seen that DeepFace outperformed across all metrics; therefore, the DeepFace results are used to compare with the results of this study.

Over the years, YOLO models have become quite popular and essential in the real-time detection of objects, with a rollout of a new model to deal with the limitations of predeceasing versions happening every now and so often. At the time of writing, the latest version of the model was YOLOv8; considering that, Terven and Cordova-Esparza, performed a comprehensive review study of the various YOLO models [14]. The study highlights the changes made to the model, such as new network backbones, a better technique to augment data, and different optimization techniques. The study also points out that from YOLOv4, there were slight improvements in model accuracy without affecting the real-time performance of the models but with an increase in computational costs. Therefore, YOLOv4 is used in this work.

In a research study to develop and optimize a weapon detection system, Ahmed et al. sought to improve real-time weapon detection and explored large-scale deployment [15]. The study highlights key areas of large-scale deployment and challenges that arise from cloud deployment hardware limitations and uses that information not only to improve the accuracy and performance of existing weapon detection models and improve the number of frames per second for real-time deployment but also to compare the performance of the models on different computing devices. The methodology emphasizes the importance of datasets and how they affect algorithms. The dataset used for training and testing consisted of 8327 labeled images split into training and testing datasets. To improve the mAP model, the datasets were broken into two categories, pistol and non-pistol, which helped reduce false negatives and false positives. The model used was YOLOv4 and TensorRT to optimize the model into YOLOv4-Scaled for network optimization in edge-computing with comparison results favoring the YOLOv4-Scaled model. In future works, Ahmed et al. mention using body poses to improve the system [15]. This article has aided my research project by giving me insight into developing systems for large-scale deployment optimization techniques and a successful systematic approach to developing a weapon detection system using YOLO.

1.2 State-of-the-art review

This review discusses the state-of-the-art technologies in emotion recognition and weapons detection systems, examining their applications in enhancing safety for ridesharing passengers and drivers.

1.2.1 Emotion recognition technologies

Emotion recognition refers to the technological capability to identify human emotions through various means, including facial expressions, voice intonation, and physiological signals. In ridesharing, monitoring driver emotions can significantly impact driving safety.

1.2.1.1 Technological advances

Facial expression analysis: Machine learning algorithms, particularly convolutional neural networks (CNNs), are increasingly used to analyze facial expressions captured *via* in-vehicle cameras. Cai et al. reported that these systems can detect emotions such as anger or frustration, which are known to correlate with risky driving behaviours [16].

Behavioural analysis: Recent studies have proposed using vehicle dynamics—such as heavy braking or sharp turns, as indicators of driver emotional states. This non-intrusive approach leverages existing vehicle sensors to infer emotional states without compromising privacy [17].

1.2.1.2 Impact on safety

Proactive interventions: Emotion recognition technologies can provide real-time feedback to drivers when negative emotions are detected. For instance, alerts could prompt drivers to take breaks or engage in calming activities as indicated by Li et al. and Liu and Wang [18, 19].

Passenger safety assurance: Passengers may feel safer knowing that their driver's emotional state is being monitored, potentially reducing anxiety during rides [16, 20].

1.2.2 Weapons detection systems

The integration of weapons detection systems in ridesharing vehicles is crucial for enhancing security. These systems employ advanced technologies to identify firearms and other weapons before they can pose a threat.

1.2.2.1 Detection technologies

Computer vision systems: Technologies like ZeroEyes utilize existing surveillance camera networks to detect brandished firearms through AI-driven image analysis. This system alerts security personnel within seconds of detection, facilitating rapid response [21].

Millimetre wave technology: This technology detects concealed weapons by analyzing the density of objects without requiring individuals to stop or empty their pockets. It has been successfully piloted in various public transit environments [22].

1.2.2.2 Implementation in ridesharing

Real-time monitoring: By integrating weapons detection systems into ridesharing vehicles, companies can ensure immediate alerts if a weapon is detected, allowing for swift action from law enforcement [23].

Enhancing passenger security: The presence of weapons detection technology may deter individuals from carrying weapons into ridesharing vehicles, thereby enhancing overall safety [24].

1.2.3 Challenges and considerations

Emotion recognition and weapons detection raise ethical issues around privacy and consent, particularly in personal or semi-private spaces like vehicles. Balancing safety with ethical considerations is crucial. Privacy-preserving models, which anonymize data post-analysis or ensure minimal data retention, are being researched as solutions to these concerns.

Privacy concerns: Continuous monitoring raises ethical questions regarding passenger privacy. Clear policies must be established to protect user data while ensuring safety [12, 18].

Technological limitations: Current emotion recognition systems may struggle with accuracy under varying conditions such as poor lighting or occlusions [16, 25].

1.2.4 Ongoing research is necessary to enhance these technologies

Gaining public trust in surveillance technologies is critical. Transparent communication about the purpose and benefits of these systems can help alleviate concerns [16, 20].

The integration of emotion recognition and weapons detection systems within ridesharing services presents a promising strategy for enhancing safety for both drivers and passengers. By leveraging advanced technologies, ridesharing companies can create a more secure environment that addresses immediate risks while fostering trust among users. Future research should focus on improving system accuracy, addressing privacy concerns, and ensuring user acceptance to fully realize the potential benefits of these innovations.

From the literature, we now understand the related work and tools available to achieve our proposed work. Notable limitations and gaps are seen in [6] with the proposed dashcams and watchdog networks solution. The solution requires the live footage to be made accessible through social media platforms and observed by strangers (watchdog network); this limitation is that passengers might not feel comfortable being "watched" by strangers. Their trip was made public, and my proposed work takes care of this by removing the need to have people monitoring the trip and substituting it with a system that does both fear recognition and handgun detection. The work done by Ruiz-Santaquiteria et al. only utilizes individual images [11].

The escalating incidents of harassment, gender-based violence, and driver vulnerability in cash-based transactions require an innovative approach to safety measures within ride-share services. The traditional safety features have limitations in critical situations, emphasizing the urgency for a more robust and responsive system to ensure the safety and well-being of ride-share users. This chapter aims to contribute to passenger and driver safety of ride-share users by developing an image processing system that uses machine learning techniques for emotion recognition and weapon detection. The chapter contributes to ridesharing safety by expanding image processing techniques beyond driver monitoring implemented in related literature and including passengers. This study implements an image processing system *via* two Streamlit web apps, incorporating facial emotion recognition (FER) and handgun detection.

2. Methodology

The processing the emotion recognition and weapons detection system's information and data involves a sequence of computational steps that transform raw inputs, such as facial images or object data, into actionable outputs, such as detected emotions or weapon alerts. Below, the methods are broken down into their most basic elements, highlighting the core principles and technologies.

Basic workflow of the system

1. Data collection: Captures raw video and audio inputs.

2. Preprocessing: Prepares the data by isolating faces or objects and enhancing quality.

3. Feature extraction: Identifies key patterns like facial landmarks or object textures.

4. Model analysis: Classifies extracted features into emotions or objects using trained AI models.

5. Decision system: Triggers alerts based on the classification results.

6. Output delivery: Provides real-time feedback *via* a user interface and alert systems.

2.1 Data acquisition

Input sources: Cameras capture real-time video streams of the vehicle's interior.

Data types: Image Data - Captures facial expressions and objects.

2.2 Preprocessing

Preprocessing prepares raw data for analysis by improving its quality and suitability for recognition algorithms.

Image preprocessing: Convert video frames to grayscale. Resize images to match model input dimensions (e.g. 48 × 48 pixels for FER2013-based models). Normalize pixel values (e.g. scale from 0 to 255 to 0–1) to improve computational efficiency.

Face detection: Apply algorithms like Haar Cascades or deep learning-based methods (e.g. Multi-task Cascaded Convolutional Networks (MTCNN)) to isolate faces from the video frames for further emotion analysis.

Object detection preparation: Use bounding box algorithms to focus on potential regions containing objects for weapon detection.

2.3 Feature extraction

Feature extraction identifies and highlights relevant patterns in the data, which are then fed into machine learning models.

Facial features for emotion recognition: Extract landmarks (e.g. eye, nose, and mouth positions) to aid in identifying facial expressions. Input pixel values into CNN layers to identify high-level features corresponding to emotions.

Object features for weapon detection: Input object region data into YOLOv4 or similar models to identify characteristic shapes and textures of weapons.

2.4 Model processing

This stage involves using trained machine learning models to process the extracted features and generate outputs.

Emotion recognition: CNNs: Convolutional Neural Networks process image data through layers of filters, pooling, and fully connected layers to classify emotions (e.g. angry, fear, and happy).

Weapon detection: YOLOv4 - Processes regions of interest to detect objects in the frame, assigning probabilities to classify them as weapons or non-weapons.

2.5 Decision making

After processing, the system interprets the model's outputs to make safety-related decisions.

Emotion analysis: High-confidence detection of negative emotions (e.g. fear and anger) triggers an alert system.

Weapon detection: Positive identification of a weapon triggers a higher-priority alert, which may involve emergency escalation.

2.6 Alert system

The processed results are communicated to relevant stakeholders *via* a user interface.

Visualization: The Streamlit GUI displays real-time results, such as the detected emotion ("Fear") or object ("Handgun").

Notifications: If a safety risk is detected (e.g. a weapon or heightened emotion), the system sends alerts to the driver and passenger (in-app notifications), and the ridesharing company's support or emergency team.

This chapter presents a two-part danger detection collaborative safety tool: a GUI web app that ride-share companies like Uber and Bolt can use during trips. The two detection methods employed are emotion detection and handgun detection. These two detection methods are integrated using a Streamlit GUI. An alarm is triggered upon detecting the respective threats, and an image is saved. This section explains the integrated threat identification and response approach. **Figure 1** is a flowchart outlining the important stages of the overall project.

Figure 1 presents an overview of the research design and methodology for integrating emotion recognition and weapons detection systems in a ridesharing safety framework. The design includes two main detection pathways—one focused on weapons (e.g. handguns) and the other on emotional states (e.g. fear). These options are outlined as follows:

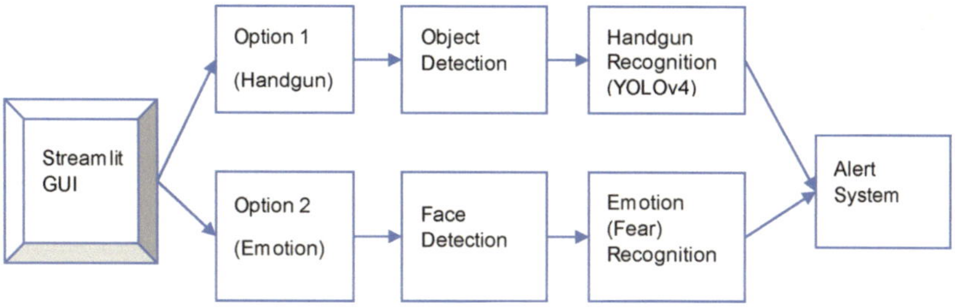

Figure 1.
Overview of research design/methodology.

Object detection and face detection: Both pathways begin with object detection and face detection processes. Object detection identifies items within the vehicle environment, while face detection isolates face for emotional analysis.

Handgun recognition (YOLOv4): For weapon detection, YOLOv4 (you only look once version 4) is used to recognize handguns specifically. This model is selected for its real-time detection capabilities, allowing the system to alert operators immediately if a weapon is identified.

Emotion recognition (fear) using sequential CNN: For emotion detection, a sequential convolutional neural network (CNN) is employed to recognize specific emotional cues, particularly fear. This model processes facial expressions to identify emotional states that could indicate distress or potential conflict.

Streamlit GUI and alert system: Both detection pathways feed into a Streamlit GUI, which serves as a user interface, displaying real-time outputs from the detection systems. If either a weapon or an emotional cue like fear is detected, an alert system is triggered to notify the ridesharing support team, enabling prompt intervention.

This dual-path approach allows the system to detect both physical threats and high-risk emotional states, enhancing ridesharing safety through a layered, proactive monitoring solution (**Figure 1**).

2.7 Dataset collection

For this chapter, both datasets are open data found on the Kaggle platform. For training and testing the Emotion Recognition model, the FER2013 dataset introduced by Dumitru et al. was utilized [26]. In contrast, the "Handgun Detection" dataset provided by the University of Granada [10] was used for the Handgun detection model. Exploration and preprocessing were conducted on the dataset.

2.7.1 FER2013 dataset

The FER2013 dataset (Facial Expression Recognition 2013) is a widely used benchmark for emotion recognition tasks, particularly in applications involving computer vision and machine learning. It was originally introduced for the ICML (International Conference on Machine Learning) 2013 competition and has since become a standard dataset for training and evaluating models designed to recognize human emotions through facial expressions.

The FER2013 Dataset consists of 32,568, 48 × 48 grayscale images of faces displaying seven different emotions, that is, anger, disgust, fear, happiness, neutrality, sadness, and surprise which is described in the **Figure 1** as Option 2 (Emotion). From the 32,568 images, the dataset is further split into 28,709 Train images and 3859 Test images.

FER2013 is extensively used for training convolutional neural networks (CNNs) in facial expression recognition tasks. The dataset is particularly valuable for deep learning models like CNNs due to its size and diversity of facial expressions. Many modern emotion recognition systems, including those for customer service, healthcare, and security applications, use FER2013 to develop real-time recognition capabilities.

For ridesharing safety applications, FER2013 can be instrumental in training models to detect emotions like "Fear" or "Angry," which are associated with potential risks. By recognizing these emotions in real time, systems can alert operators or support teams to de-escalate situations before conflicts arise. Its range of emotions makes FER2013 particularly useful for recognizing distress or aggression in confined environments, enhancing safety monitoring in scenarios like ridesharing or public transportation.

By exploring the dataset folders, it was noticed that both the train and test sets were imbalanced, as seen in **Figure 2**. Using imbalanced datasets to train models is not good practice as it introduces a bias towards the dominant class in the dataset.

This study implemented data augmentation to balance the datasets and reduce some folders. In addition to balancing out the datasets, sampling was also introduced. Sampling uses a fraction of the total dataset as opposed to the entire dataset due to the high computation costs of using the whole dataset. The final balanced datasets comprised 28,000 training images and 4655 testing images, using an 85/15 train-test split. Each training emotion has 4000 images, while each test emotion hosts 665 images.

The balanced datasets were then normalized by dividing each pixel value by the max pixel value, 255. Normalization is a crucial preprocessing step as it ensures all inputs follow a standard distribution, ensuring the gradient calculations computed in the model network remain consistent, helping the model train better and faster. The final preprocessing step was resizing the images to match the required input size for each model.

2.7.2 Handgun detection dataset

The Handgun Detection Dataset comprises 2986, 416 × 416 images [10]. The images are wide-ranging, all containing different handguns/pistols, images taken at various angles, handguns in hand, cartoon images, and staged studio-quality pictures of guns. The dataset also had its labels, but this study did not use the labels

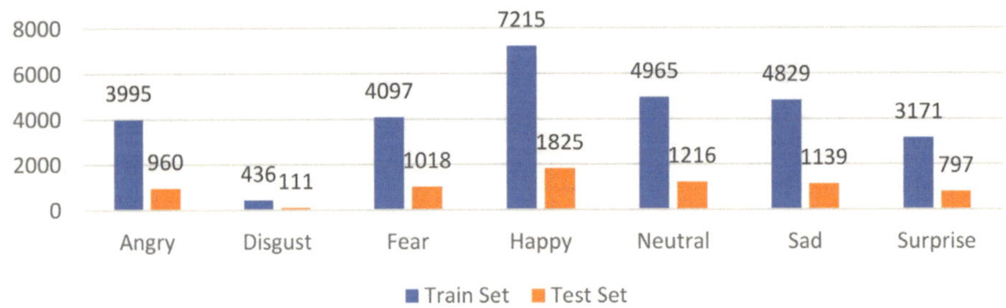

Figure 2.
Number of images for each emotion in the train and test sets.

(gun position in the image) provided and opted to create new labels for each separate image instead of having it all stored in one file.

This study made use of LabelImg to create new labels. LabelImg is a user-friendly tool designed to annotate object bounding boxes in images. It supports the YOLO text format. It is a lightweight, open-source software developed by Tzutalin in 2015 that utilizes graphs to label imagesLabelling individual images is quite time-consuming; therefore, this study randomly selected and labelled a sample of 265 images. The minimum number of images needed for object detection, according to Lee 2021, ranges between 150 and 500 [27].

With the model needing labeled image annotations, traditional data augmentation techniques would prove to be troublesome; therefore, to increase the dataset, several duplications of images and the respective annotations were made. Ultimately, the dataset consisted of 925 images and 925 annotations. A 70/30 train-test split was then employed, resulting in 650 training images and 275 testing images.

Since the object detection was not a multiclass classification, with handgun being the only class being identified, there was no need to balance out the dataset.

2.8 Modeling

To achieve robust facial emotion recognition, this study trained/used and compared three distinct models: the pre-trained model offered by the lightweight DeepFace python library, a sequential CNN, and MobileNetv2 using transfer learning. The selection of DeepFace was due to it outperforming its lightweight counterparts, as seen in [13], as well as it being a lightweight open-source library for facial recognition comprising several state-of-the-art AI models specialized in facial recognition, such as VGG-Face, Facenet, OpenFace, DeepFace, Dlib, and ArcFace. Notable lightweight absentees are the Sequential and MobileNetv2 models; hence, they were the other models trained and tested in this study. Additionally, utilizing the YOLOv4 model is a cornerstone for object detection. Its cutting-edge architecture and capabilities are ideal for detecting objects, including handguns.

2.8.1 Facial emotion recognition

Figure 3 gives a brief overview of the critical stages in the facial emotion recognition model in the form of a flow chart. The key stages are further discussed below.

Figure 3 provides a concise overview of the facial emotion recognition process, which involves the following sequential steps:

1. *Image acquisition:* This is the initial step where images are captured using cameras installed in the ridesharing vehicle or other monitoring systems. These images serve as the raw data for processing. Live image footage of different facial emotions was acquired in this study *via* the laptop webcam.

Figure 3.
Overview of facial emotion recognition.

2. *Face detection*: The system isolates faces from the captured images using algorithms like Haar Cascades, Multi-task Cascaded Convolutional Networks (MTCNN), or other deep learning-based methods. This step focuses only on the regions of interest (the faces) while ignoring other parts of the image.

3. *Face recognition*: After detecting a face, the system analyzes it to identify specific emotional features. Convolutional Neural Networks (CNNs) or similar machine learning models are typically employed here to classify the emotion (e.g. anger, fear, and happiness) based on the detected facial features.

4. *Safe and alert*: Depending on the recognized emotion, the system decides. The image or video frame containing the detected handgun is stored in a secure database for record-keeping and potential evidence. Also, an immediate alert is triggered to notify relevant stakeholders, such as the ridesharing company's security team or law enforcement, ensuring timely intervention.

Before any facial emotions are detected and inferences are made, facial detection is done on the acquired image to ensure a face. This study used the OpenCV Haar-cascade classifier for face detection by downloading and loading it into the haarcascade_frontalface_default.xml file. Haar Cascades is an object detection algorithm for finding faces in pictures or real-time video analysis. The algorithm uses edge and line detection and works by using "positive images" with faces and "negative images "that do not have faces. According to Behera 2020, the algorithm trains these positive and negative images to identify if images have a face in them [28]. After the face is detected, a bounding box is drawn around the face.

2.8.1.1 DeepFace

Since DeepFace is a pre-trained model, all that was done to utilize it was to import the DeepFace library from the deepface package. DeepFace has an inbuilt function called "analyze" which offers information on the detected face from the acquired image, such as the different possible emotions as well as the dominant emotion, age, gender, and various possible races as well as the dominant race, an example of this can be seen in **Figure 4**. This study focuses only on the predicted dominant emotion, which is then printed in the bounding box created during the face detection step.

Figure 4 illustrates the functionality of the DeepFace framework for analyzing facial data to recognize emotions. DeepFace is an advanced facial analysis tool that utilizes deep learning to process and interpret facial features.

Face detection: The system first identifies and isolates faces from the input image or video stream. This step ensures that only relevant facial regions are analyzed.

Feature extraction: Using a deep learning model, DeepFace extracts key facial features such as the eyes, nose, mouth, and other distinguishing elements. These features are converted into numerical representations (embeddings) for processing.

Emotion recognition: The extracted features are processed to classify the individual's emotional state. DeepFace typically recognizes a range of emotions, such as happiness, sadness, anger, surprise, and fear, using pre-trained neural networks.

Output generation: The results of the analysis are displayed, indicating the detected emotion(s) with corresponding confidence levels. These outputs can be visualized through a graphical user interface or logged for further analysis.

```
{'emotion': {'angry': 0.7797068916261196,
  'disgust': 1.4739168818778126e-05,
  'fear': 68.81503462791443,
  'happy': 0.03573041467461735,
  'sad': 13.606315851211548,
  'surprise': 9.098193049430847,
  'neutral': 7.665000110864639},
 'dominant_emotion': 'fear',
 'region': {'x': 279, 'y': 73, 'w': 96, 'h': 96},
 'age': 23,
 'gender': 'Woman',
 'race': {'asian': 0.07142509879319682,
  'indian': 0.031692231137672826,
  'black': 0.0011916193614634422,
  'white': 93.0402525615391,
  'middle eastern': 3.355551483565392,
  'latino hispanic': 3.4998833686252393},
 'dominant_race': 'white'}
```


Figure 4.
Example of DeepFace analysis function.

2.8.1.2 MobileNetV2

MobileNetV2 is a lightweight convolutional neural network model built for image classification. It aims to excel in performance when deployed on mobile and edge-computing devices; hence, it was selected. It adopts an inverted residual structure, employing residual connections within bottleneck layers. According to Sandler et al. the intermediate expansion layer utilizes lightweight depth-wise convolutions to filter features, providing non-linearity [29]. The model is downloaded from tensorflow keras.

Keras has a pre-trained MobileNetV2 model; therefore, transfer learning was implemented in this study. Transfer learning refers to leveraging knowledge/features gained from one task and applying it to a different but related task [30]. MobileNetV2 is said to be trained "on more than a million images from the ImageNet database" [31].

Due to the limitation on the computation resources, MobileNetV2 only allowed 1400 images to be trained. Therefore, the dataset was reduced to 1400 train images (200 per emotion) and 350 test images (50 per emotion).

MobileNetV2 takes the input of image size 224 × 224, 3. Therefore, the images were resized to fit this required input size.

Several new layers were added to the pre-existing model; two dense layers with 128 and 64 units perform a linear operation on the input from the previous layers, using ReLU activation functions, which introduce non-linearity. Dropout layers with a dropout rate of 0.2 after each dense layer to prevent overfitting. A final dense layer with seven units since it is a classification task with seven classes (seven emotions) using a softmax activation function to output probabilities for each class. Softmax was chosen as it is suitable for a multiclass problem.

Figure 5a and **b** shows the training of the MobileNetV2 model. The model achieved high training accuracy, but as seen by the validation accuracy line, overfitting occurred despite the dropout layers as the validation accuracy continuously decreased (**Figure 5a**). **Figure 5b** illustrates the training and validation performance of the MobileNetV2 model during the training process. This comparison highlights how well the model generalizes to unseen data after being trained on a specific dataset.

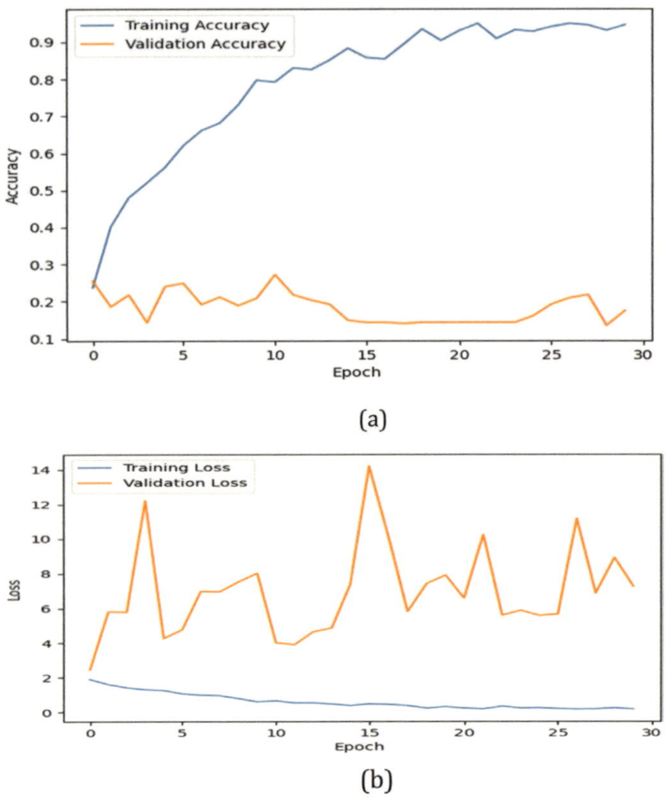

Figure 5.
Training accuracy versus test accuracy of MobileNetV2 model. (a) Model accuracy (training accuracy vs. validation accuracy). (b) Model loss (training loss vs. validation loss).

2.8.1.3 Sequential CNN

The model takes an input of size 48 × 48 with three channels, which is already the size of our dataset; however, images were resized to the desired input to prevent any unnoticed incorrect input size.

The architecture employed to build the Sequential model:

- Conv2D layer: Constructs a convolutional layer with a given number of filters (32, 64, or 128), kernel size (3, 3), ReLU activation function, 'same' padding to preserve spatial dimensions, and an input shape for the first layer. The filters increase from 32 to 64 to 128, gradually capturing more complex features.

- BatchNormalization layer: Normalizes the previous layer's activations, allowing faster convergence during training and minimizing overfitting.

- MaxPooling2D layer: Down samples the spatial dimensions by taking the maximum value inside a defined window (2 × 2).

- Dropout layer: The dropout layer prevents overfitting by setting a certain percentage of the input to 0, preventing the model from memorizing the data and enabling learning patterns. In this case, 25% of the input is set to 0.

- Flatten layer: Flattens the previous layers' output into a 1D array, making it ready to input the dense layers.

- Dense layer: Fully linked layers that use ReLU activation to learn high-level information and make predictions. The final dense layer includes the 'softmax' activation function typically used for multiclass classification.

Figure 6a–d shows the training of the sequential CNN model. We trained the model twice, 30 epochs each; however, the model was first trained for 18 and 16 epochs on the second attempt due to early stopping implementation. We were able to increase the accuracy from 59–61%. The model trained relatively well; there are still signs of overfitting; however, the validation accuracy curve seemed to follow an increasing trend despite the validation loss not decreasing.

2.8.2 Handgun recognition

Figure 7 gives a brief overview of the key stages in the facial emotion recognition model in the form of a flow chart. The key stages are further discussed below.

Figure 7 outlines a streamlined process for handgun detection and alert generation, composed of three core steps, shown below. This workflow demonstrates a robust mechanism for enhancing safety in real-time environments by integrating AI-based object detection with actionable alert systems. The simplicity of the process allows for quick deployment and reliable operation in high-risk scenarios.

1. *Image acquisition*: This step involves capturing real-time video or image data using cameras installed in the monitoring environment, such as ridesharing vehicles or surveillance areas. The acquired images serve as the raw input for further analysis, ensuring continuous monitoring to detect potential threats.

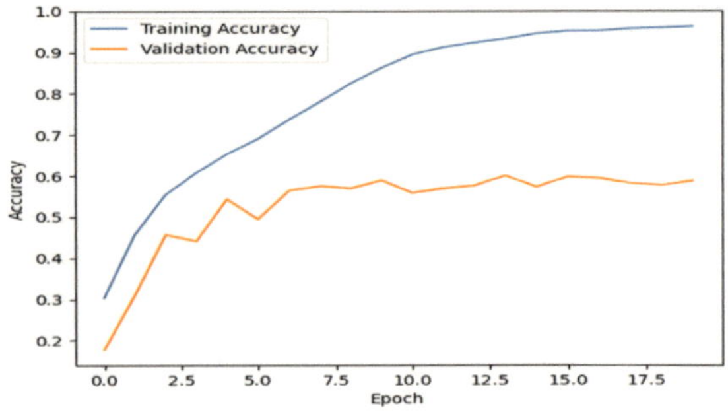

(a)

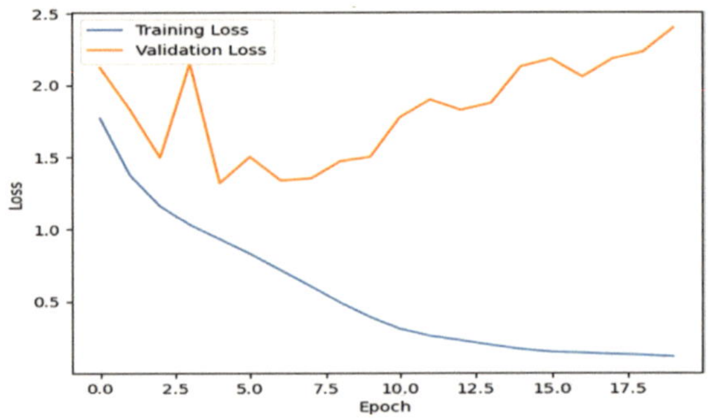

(b)

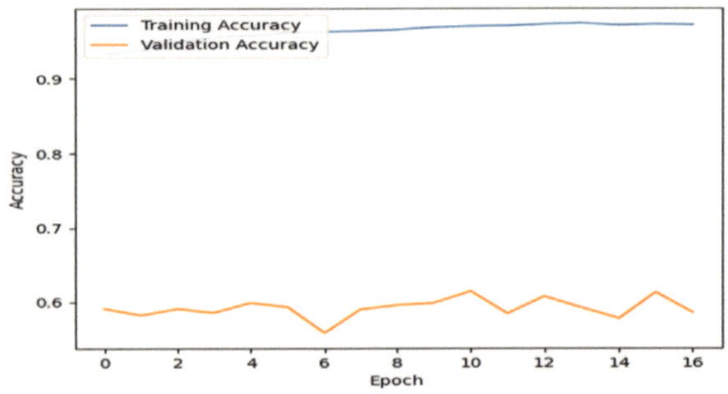

(c)

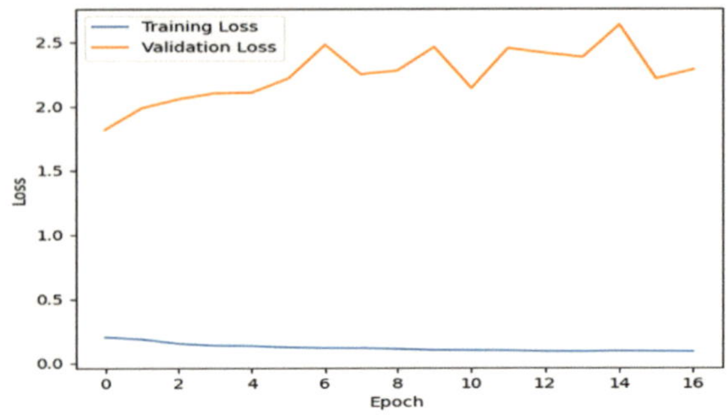

(d)

Figure 6.
Training accuracy versus test accuracy of sequential CNN model for epochs 18 and 16, respectively. (a) Model accuracy (training accuracy vs. validation accuracy) Epoch 18. (b) Model loss (training loss vs. validation loss) Epoch 18. (c) Model accuracy (training accuracy vs. validation accuracy) Epoch 16. (d) Model loss (training loss vs. validation loss) Epoch 16.

DOI: http://dx.doi.org/10.5772/intechopen.1008540

Figure 7.
Key stages of handgun detection.

Live image footage of different handguns in hand was acquired in this study *via* the laptop webcam.

2. *Object detection (handgun)*: Advanced object detection algorithms, such as YOLOv4, process the captured images to identify handguns. This involves isolating regions of interest within the image, extracting relevant features, and classifying them as either "handgun" or "non-handgun" objects. The system uses deep

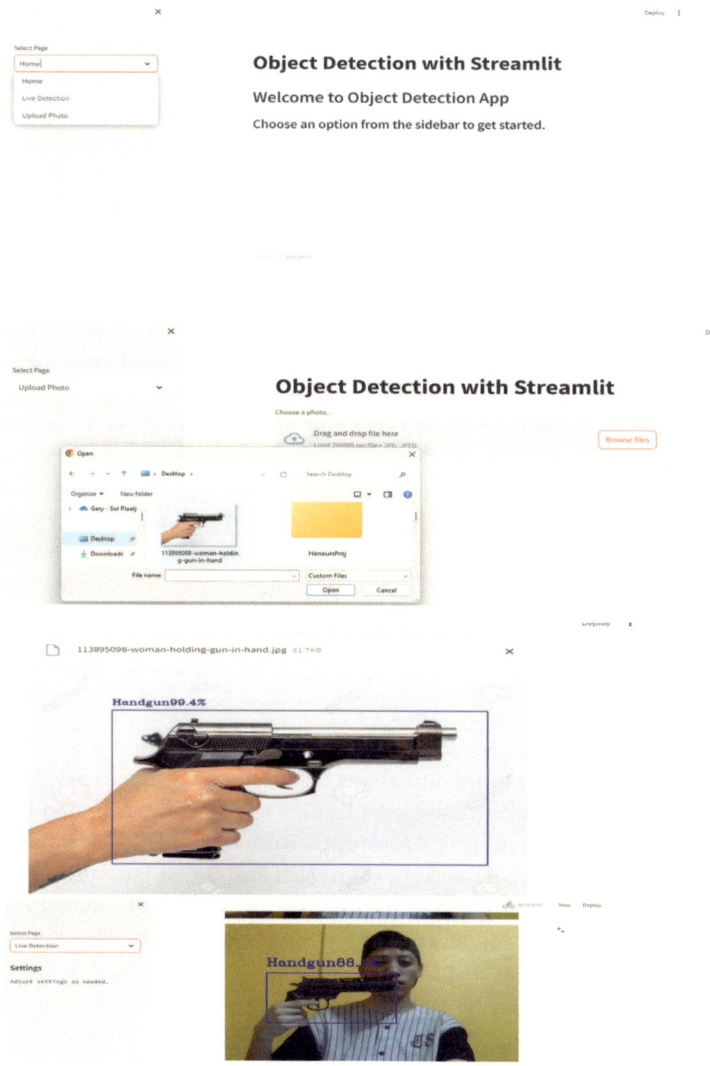

Figure 8.
Object detection via Streamlit.

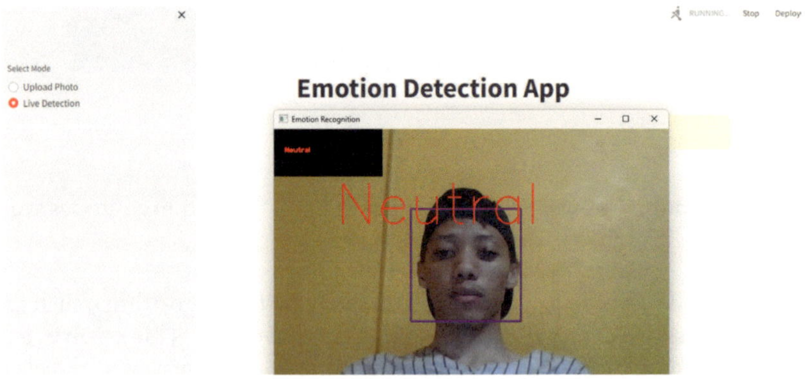

Figure 9.
Emotion detection via Streamlit.

learning models trained on large datasets of weapon images to achieve high accuracy and real-time performance. According to Korlakunta, YOLOv4 is the fourth iteration of the YOLO algorithm, which introduced a single-stage, end-to-end approach to object detection [32]. Unlike standard two-stage detectors, YOLOv4 processes the entire image in a single pass, making it highly efficient. It delivers cutting-edge precision by combining advanced techniques such as a strong backbone network, a feature pyramid network, and numerous detecting heads [32]. The YOLOv4 used Darknet, an open-source neural network framework, to perform object detection in this study. The Darknet repo is available on the AlexeyAB GitHub account [33]. In addition to the Darknet, YOLOv4 uses a configured file and pre-trained weights, available on the DominykasJurkus GitHub account [34]. My thanks and acknowledgement go out to these users. The pre-trained weights serve as a starting point for training.

3. *Save and alert*: When a handgun is detected, the system performs two actions: The image or video frame containing the detected handgun is stored in a secure database for record-keeping and potential evidence. Also, an immediate alert is triggered to notify relevant stakeholders, such as the ridesharing company's security team or law enforcement, ensuring timely intervention.

2.8.2.1 Graphical user interface

This study used Streamlit. Mhadhbi states for GUI, "Streamlit is a free and open-source framework to rapidly build and share beautiful machine learning and data science web apps [35]. It is a Python-based library specifically designed for machine learning engineers." In this study, two web apps were created, 1 for Handgun detection and another for Emotion detection. **Figures 8** and **9** display the home pages of the respective web apps. Within the GUI, the saved weights and models generated from the notebooks used for model training are loaded and integrated.

3. Results and discussion

To measure and compare the models' performances, the selected metrics are Accuracy, Precision, Recall, and F1-score.

3.1 Facial emotion recognition

The observed results of the model are displayed in **Table 1**:

Given the small dataset, it was expected that the MobileNetV2 would not perform well on the testing set despite performing well on the trained dataset. The best all-around performance on the selected metrics was the sequential CNN.

3.2 Handgun recognition using YOLOv4

The following statistics and metrics were taken at a confidence threshold of 0.25. It took a total time of 5 seconds to perform all the detections.

3.3 Comparative analysis

- DeepFace versus MobileNetV2:
    - Across all selected metrics, DeepFace significantly outperforms MobileNetV2.
    - This significant difference indicates that DeepFace is more adept at recognizing emotions in the dataset than MobileNetV2.
- Sequential CNN versus DeepFace:
    - Our sequential CNN outperforms DeepFace across all metrics.
    - This highlights that it not only classifies instances more accurately overall, but sequential CNN demonstrates a higher accuracy in identifying true positives among its positive predictions.
    - It is also worth noting that DeepFace is trained on the larger, imbalanced dataset, whereas sequential CNN is trained on a balanced sample dataset.
- Sequential CNN versus MobileNetV2:
    - Sequential CNN displays a notably better performance across all metrics compared to MobileNetV2.
    - This significant difference in performance demonstrates Sequential CNN's suitability and effectiveness for emotion recognition on the FER2013 dataset compared to MobileNetV2.

Model	Accuracy	Precision	Recall	F1-score
DeepFace	0.554	0.555	0.522	0.535
MobileNetV2	0.174	0.048	0.174	0.073
Sequential CNN	0.614	0.642	0.615	0.624

Table 1.
Model comparison.

- Classifications
    - o With an accuracy of 61%, there will be some misclassifications, as seen in **Figures 10** and **11**; however, most of these misclassifications also occurred within the other emotions and not just the fear emotion and, therefore, should not hinder the overall performance of the app to significant effect.

Sequential CNN is the most suitable choice for emotion recognition on FER2013 due to its superior performance across all metrics. MobileNetV2's inferior performance across all metrics suggests it might not be the best choice for this study. DeepFace scored more than 50% across all metrics; however, it could not outperform our sequential CNN, with our sequential CNN model scoring more than 60% across

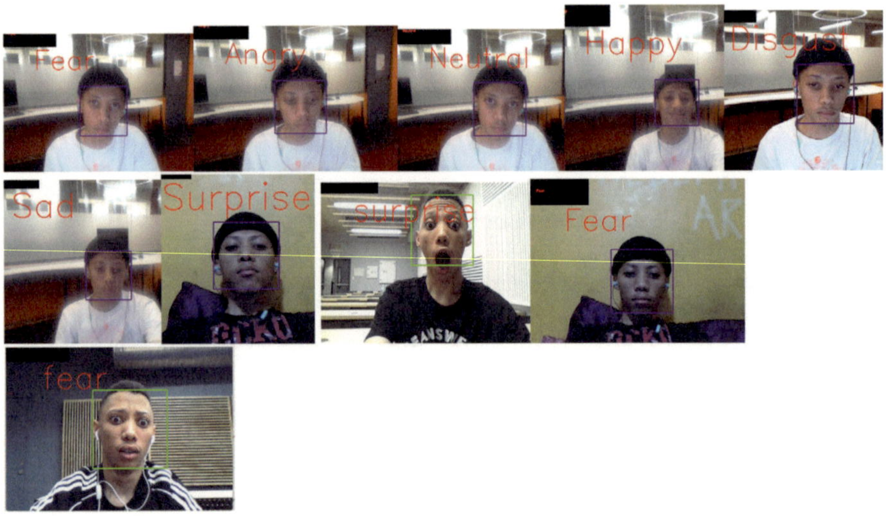

Figure 10.
Emotion classifications of sequential CNN.

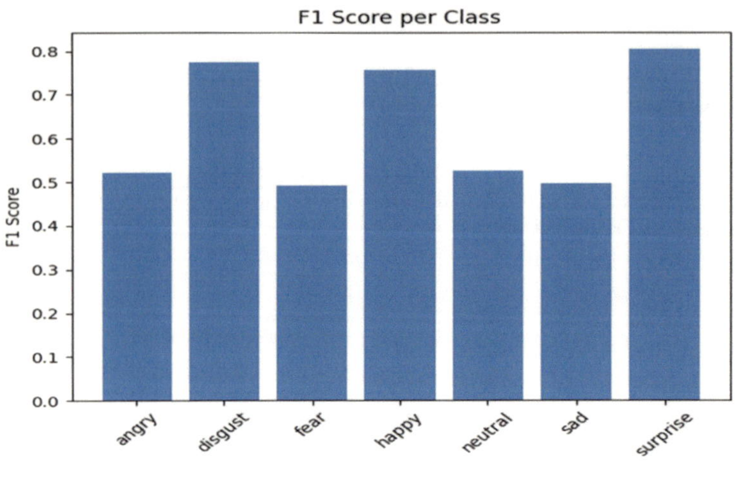

Figure 11.
F1-score per emotion.

Statistic	Number of handgun images
True positive	228
False positive	32
False negative	84

Table 2.
Statistics for handgun detection.

Metric	YOLOv4 performance
Accuracy	0.80
Precision	0.88
Recall	0.73
F1-Score	0.80

Table 3.
Performance measure of YOLOv4 on handgun detection.

all metrics, and this could be due to the imbalanced FER2013 dataset DeepFace was trained on. Sequential CNN is the clear, better-performing model used to recognize the different emotions in the Streamlit App.

Table 2 shows that the model is better at correctly identifying images that contain handguns (TP = 228) than missing them (FN = 84). Given the number of false negatives, there is room for improvement in recognizing handguns in the images that have not been detected.

The model obtains a good false positive score (FP = 32); however, reducing false positives further could enhance the precision of the model, ensuring that fewer images are incorrectly labeled as containing handguns when they are not.

Table 3 suggests that the overall performance is satisfactory, obtaining an accuracy of 0.80. The balance between precision and recall seems slightly skewed towards precision (0.88 vs. 0.73), which suggests that the model potentially missed some actual handguns.

The screenshots of the model in action seen in **Figure 12** show that the model accurately detected the handguns with high accuracy, given different types of handguns, different angles, and different backgrounds.

4. Conclusion

This chapter successfully designed a collaborative safety tool in two separate Streamlit GUI (handgun detection and facial emotion recognition) for passenger and driver safety within ride-share services by developing an image processing system employing machine learning for emotion recognition and weapon detection. The objectives outlined were to detect fear from passengers' facial expressions, identify handguns within a vehicle, and promptly alert in the presence of danger if the two threats were detected. In addition to the objectives at the start of the study, research questions were formulated: What method can be used to detect passengers' facial expressions with at least 60% accuracy? What method can be used to detect a handgun in a vehicle? What approach can be used as a suitable alert system for either fear/handgun detection?

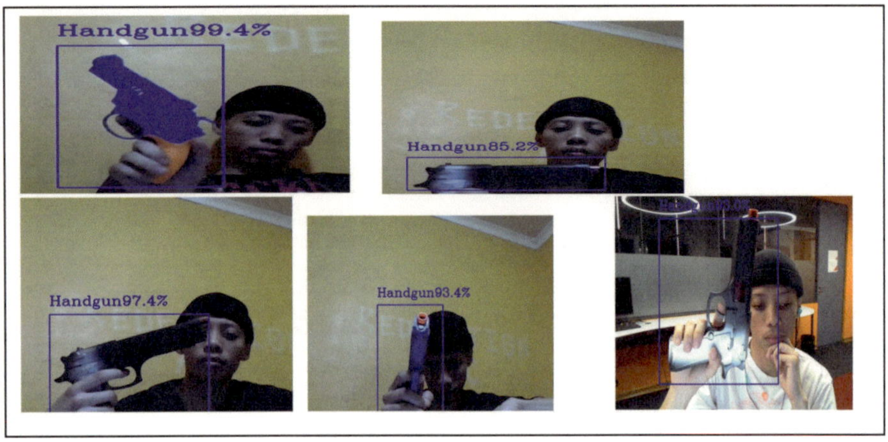

Figure 12.
Handgun detections.

The comparative analysis of DeepFace, MobileNetV2, and sequential CNN on the FER2013 dataset proved it possible to recognize emotions from facial expressions. In addition, we also answered the research question of which model/method can be used to achieve at least 60% accuracy, with the sequential CNN model reaching an accuracy of 61%, which is reasonable given an ongoing research field. The sequential CNN exhibited commendable precision, recall, and F1-score, making it the preferred model due to its balanced performance and adaptability to our application. For Handgun Detection, the YOLOv4 model proved to be a great choice as it demonstrated proficient identification of images containing handguns, achieving an accuracy of 0.80. Saving the frames with either fear/handgun detected and sounding an alarm is a suitable alert system, but it can be improved.

This chapter contributes to the evolving landscape of safety technologies in ride-share services. The findings are not meant to replace existing safety measures but emphasize the potential of machine learning and image processing in mitigating safety concerns. The study hopefully acts as a steppingstone towards enhancing passenger and driver safety and highlights the importance of ongoing research in this domain. In conclusion, leveraging sequential CNN for emotion recognition and YOLOv4 for handgun detection, the developed image processing system represents a step in the right direction and one more "gear shift" closer to safer trips.

The implications of our work are significant for ride-share safety, laying the groundwork for advanced safety measures. Given certain limitations, such as computational resources, further improvements in the models are possible, especially regarding emotion recognition. A larger and better dataset can be used to improve accuracy. Transformers are rising in computer vision use cases and could produce better results. User feedback could also be implemented to gauge the general perception of the model and whether this study would be a welcomed addition in the attempt to make ridesharing trips safer. Concerning weapon detection, the dataset can be increased. It could include more weapons to expand its use beyond detecting handguns, which will play a pivotal role in ensuring a safer and more secure environment for all ride-share users.

Future work will encompass a diverse range of test scenarios. Additionally, input from psychologists and security specialists will be sought to design a system that emphasizes core principles of psychological security while addressing key aspects of safety. The hypotheses underpinning our approach and the implications of the proposed method will be comprehensively detailed.

Acknowledgements

The authors thanked Sol Plaatje University for infrastructural support.

Conflict of interest

The authors declare no conflict of interest.

Author details

Gary Leander and Ibidun Christiana Obagbuwa*
Faculty of Natural and Applied Sciences, Department of Computer Science and Information Technology, Sol Plaatje University, Kimberley, South Africa

*Address all correspondence to: ibidun.obagbuwa@spu.ac.za

References

[1] Xiao H, Li W, Zeng G, Wu Y, Xue J, Zhang J, et al. On-road driver emotion recognition using facial expression. Applied Sciences. 2022;**12**:807. DOI: 10.3390/app12020807

[2] Sukhavasi SB, Sukhavasi SB, Elleithy K, El-Sayed A, Elleithy A. A hybrid model for driver emotion detection using feature fusion approach. International Journal of Environmental Research and Public Health. 2022;**19**(5):3085

[3] Mlamla S. Over 90 000 People Sign Petition Asking Bolt to Vet Its Drivers more Efficiently. [Blog] IOL. 2022. Available from: https://www.iol.co.za/capeargus/news/over-90-000-people-sign-petition-asking-bolt-to-vet-its-drivers-more-efficiently-bac70db8-2f67-4797-b2fd-c57458e2ab9c [Accessed: April 29, 2022]

[4] Bonyhady N. Every Day, Two Sexual Assault or Misconduct Complaints Hit Uber. New South Wales, Australia: The Sydney Morning Herald [online]. Available from: https://www.smh.com.au/politics/federal/every-day-two-sexual-assault-or-misconduct-complaints-hit-uber-20211115-p59901.html; 2021 [Accessed: April 29, 2022]

[5] Mohan AJ. Uber Panic Button: On Paper for Security, but in Car Only for Show. Mumbai: The Indian Express Group. Available from: https://indianexpress.com/article/express-exclusive/uber-files-india-panic-button-8021204/; 2022 [Accessed: April 28, 2023]

[6] Chaudhry B, El-Amine S, Shakshuki E. Passenger safety in ridesharing services. Procedia Computer Science. 2018;**130**:1044-1050

[7] Vemou K, Horvath A. Facial Emotion Recognition [online]. Rue Montoyer, Brussels: European Data Protection Supervisor. Available from: https://edps.europa.eu/data-protection/our-work/publications/techdispatch/techdispatch-12021-facial-emotion-recognition_en; 2021 [Accessed: May 12, 2022]

[8] Mehendale N. Facial emotion recognition using convolutional neural networks (FERC). SN Applied Sciences. 2020;**2**(3):1-8

[9] Lim JZ, Mountstephens J, Teo J. Emotion recognition using eye-tracking: Taxonomy, review and current challenges. Sensors. 2020;**20**(8):2384

[10] Soft Computing and Intelligent Information Systems. Weapons Detection for Security and Video Surveillance. Soft Computing and Intelligent Information Systems. Online. Available from: https://sci2s.ugr.es/weapons-detection#Public%20datasets; [Accessed: July 1, 2023]

[11] Ruiz-Santaquiteria J, Velasco-Mata A, Vallez N, Deniz O, Bueno G. Improving handgun detection through a combination of visual features and body pose-based data. Pattern Recognition. 2023;**136**:109252

[12] Wang G, Ding H, Duan M, Pu Y, Yang Z, Li H. Fighting against terrorism: A real-time CCTV autonomous weapons detection based on improved YOLO v4. Digital Signal Processing. 2023;**132**:103790

[13] Sampaio EV, Lévêque L, da Silva MP, Le Callet P. Are facial expression recognition algorithms reliable in the context of interactive media? A new metric to analyse their performance. In:

EmotionIMX: Considering Emotions in Multimedia Experience (ACM IMX 2022 Workshop). Aveiro, Portugal; Jun 2022:hal-03789571

[14] Terven J, Cordova-Esparza D. A comprehensive review of YOLO: From YOLOv1 to YOLOv8 and beyond. arXiv. 2023;**5**(4):1680-1716

[15] Ahmed S, Bhatti MT, Khan MG, Lövström B, Shahid M. Development and optimization of deep learning models for weapon detection in surveillance videos. Applied Sciences. 2022;**12**(12):5772

[16] Cai Y, Li X, Li J. Emotion recognition using different sensors, emotion models, methods and datasets: A comprehensive review. Sensors. 2023;**23**(5):2455. DOI: 10.3390/s23052455

[17] Research Collection ETH Zurich. Detecting Emotion and Well-Being of Drivers under Naturalistic Condition. Zurich, Switzerland: Research Collection ETH Zurich; 2023. Available from: https://www.research-collection.ethz.ch/bitstream/handle/20.500.11850/541424/1/Dissertation_research_collection.pdf

[18] Li W, Li G, Tan R, et al. Review and perspectives on human emotion for connected automated vehicles. Automotive Innovation. 2024;**7**:4-44. DOI: 10.1007/s42154-023-00270-z

[19] Chen T, Wang M, Gong S, Zhou Y, Ran B. Connected and automated vehicle distributed control for on-ramp merging scenario: A virtual rotation approach, Transportation Research Part C: Emerging Technologies. 2021;**133**:103451. DOI: 10.1016/j.trc.2021.103451. Available from: https://www.sciencedirect.com/science/article/pii/S0968090X2100440X. ISSN 0968-090X

[20] Ahmed M, Kaur H. Emotion recognition using different sensors, emotion models, methods and datasets: A comprehensive review. Sensors. 2022;**23**(5):2455. DOI: 10.3390/s23052455

[21] Chicago Transit Authority. CTA Leverages Extensive Security Camera Network to Pilot Gun Detection Technology. Chicago: Chicago Transit Authority; 2024. Available from: https://www.transitchicago.com/cta-leverages-extensive-security-camera-network-to-pilot-gun-detection-technology/

[22] Metro Los Angeles. Metro Expands Safety Measures and Pilot Weapons Detection Technology. Los Angeles: Metro Los Angeles; 2024. Available from: https://abc7.com/post/metro-expands-safety-measures-pilot-weapons-detection-technology/

[23] CCTV Security Surveillance. AI Weapon Detection - CCTV Security Surveillance. Lake Mary, Florida: CCTV Security Surveillance; 2024. Available from: https://www.soundthinking.com/security/weapons-detection/

[24] Evolv Technology. AI Weapons Detection Startups Compete with Industry Giant in Expanding Virginia School Market. Front Royal, Virginia: Evolv Technology; 2023. Available from: https://royalexaminer.com/ai-weapons-detection-startups-compete-with-industry-giant-in-expanding-virginia-school-market/

[25] Gu Y, Liao X, Qin X, Yao Q. YouTube-GDD: A challenging gun detection dataset with rich contextual information. arXiv. preprint. 2022:arXiv:2203.04129. Available from: https://arxiv.org/abs/2203.04129

[26] Dumitru IG, Cukierski W, Bengio Y. Challenges in Representation Learning: Facial Expression Recognition Challenge. Kaggle; 2013. Available from: https://kaggle.com/competitions/

challenges-in-representation-learning-facial-expression-recognition-challenge

[27] Lee C. How Many Images Do you Need for Object Detection? Medium. Available from: https://changsin.medium.com/how-many-images-do-you-need-for-object-detection-d33185629843#8b20; 2021 [Accessed: November 2, 2023]

[28] Behera GS. Face Detection with Haar Cascade - towards Data Science. medium.com. Available from: https://towardsdatascience.com/face-detection-with-haar-cascade-727f68dafd08; 2020 [Accessed: October 20, 2022]

[29] Sandler M, Howard A, Zhu M, Zhmoginov A, Chen LC. Mobilenetv2: Inverted residuals and linear bottlenecks. In: Proceedings of the IEEE Conference on Computer Vision and Pattern Recognition. Salt Lake City, UT, USA: IEEE; 2018. pp. 4510-4520. DOI: 10.1109/CVPR.2018.00474. ISBN: 978-1-5386-6421-6

[30] Team K. Keras Documentation: Transfer Learning & Fine-Tuning. Keras [Online]. Date created: 2020/04/15; Last modified: 2023/06/25. Available from: https://keras.io/guides/transfer_learning/; [Accessed: November 1, 2023]

[31] Deep Network Designer. MobileNet-v2 Convolutional Neural Network - MATLAB. Deep Network Designer. MathWorks; 2024. Available from: https://www.mathworks.com/help/deeplearning/ref/mobilenetv2.html; [Accessed: November 2, 2023]

[32] Korlakunta K. Yolov4: A Comprehensive Guide to Object Detection Using Darknet and Opencv. medium.com. Available from: https://korlakuntasaikamal10.medium.com/yolov4-a-comprehensive-guide-to-object-detection-using-darknet-and-opencv-bcf1688f57d7; 2023 [Accessed: November 1, 2023]

[33] AlexeyAB. Darknet [Internet]. GitHub Repository. Available from: https://github.com/AlexeyAB/darknet/tree/master/src [Accessed: December 12, 2024]

[34] Jurkus D. [Internet]. GitHub Repository. Available from: https://github.com/DominykasJurkus [Accessed: December 12, 2024]

[35] mhadhbi, N. Python Tutorial: Streamlit. DataCamp [Online]. Available from: https://www.datacamp.com/tutorial/streamlit; 2021 [Accessed: November 2, 2023]

Chapter 8

Perspective Chapter: Silent Speech Interface Based on Neural Activity – A Short Review

Ming Zhang, Yuan Yuan and Shuo Zhang

Abstract

Silent Speech Interface (SSI) technology has emerged as a fascinating area of research with the potential to transform communication. This chapter presents an overview of SSI, beginning with exploration of the diverse sensing modalities employed to capture the neural and muscular signals. These include electroencephalography (EEG), surface electromyography (sEMG), and other emerging techniques. The subsequent sections detail the processing of neural signals, encompassing feature preprocessing, and a variety of recognition algorithms in the context of Silent Speech Recognition (SSR). Additionally, different voice synthesis methods are discussed. The application scenarios of SSI are examined, spanning from restoring speech capabilities for individuals with speech impairments to enhancing communication in noisy environments and enabling silent communication in private and military contexts. Despite its significant potential, SSI confronts several challenges, including bio-sensing noises, difficulties in model generalization, the absence of comprehensive evaluation standards, and concerns related to data security and social ethics. Future research directions focus on enhancing sensing accuracy, improving model performance and generalization capabilities, establishing standardized evaluation benchmarks, and addressing ethical considerations. In summary, SSI holds the promise of revolutionizing communication, yet substantial research and development efforts are required to overcome the existing obstacles.

Keywords: silent speech Interface, brain-computer Interface, voice reconstruction, non-invasive communication, neural signal processing, model generalization, data security

1. Introduction

In human-computer interaction (HCI), speech is crucial for human-machine communication, used in virtual assistants and voice-controlled devices. But traditional speech recognition systems, relying on audible sound waves, have drawbacks. They're sensitive to noise, affecting accuracy, and pose privacy risks as audible speech can be eavesdropped. Also, they are ineffective for speech-impaired individuals and impractical in noisy or security-sensitive settings.

The Silent Speech Interface (SSI) addresses this aspect by capturing non-acoustic signals related to speech production, such as electromyography (EMG), electroencephalography (EEG), and mechanical movements of speech organs. SSI aims to decode speech from these signals, enabling communication where traditional systems falter. It's more noise-resistant, enhances privacy, and aids speech-disabled individuals, promoting inclusive communication.

The use of non-acoustic signals for speech recognition dates back to the early twentieth century. Morse and O'Brien [1] first explored EMG signals for monitoring speech-related muscle activity. Later, with computational and signal processing advancements, non-audible speech recognition advanced. The introduction of surface EMG electrodes and improved techniques led to initial silent speech recognition attempts. The development of Brain-Computer Interface (BCI) technologies in the 1990s and 2000s was significant. Studies by Brumberg et al. [2] and Hueber et al. [3] examined brain implants and tongue sensors for silent speech recognition.

The 1980s and 1990s saw progress in non-invasive signal acquisition. Surface EMG was refined, and early BCI systems emerged. Researchers like Jorgensen et al. and Manabe [4, 5] explored EMG for subvocal speech recognition. The twenty-first century brought deep learning to silent speech recognition. Meltzner et al. [6] showed how deep learning enhanced sEMG-based silent speech recognition.

Silent Speech Recognition (SSR), the core of SSI, involves capturing, processing, and translating non-acoustic signals into speech. EMG-based SSR has been studied. Deng et al. [7] explored detecting and decoding muscle contractions. But it has challenges like electrode placement and noise. Advanced techniques and models like adaptive filtering and deep learning [8] have been developed.

BCI technologies like EEG and functional near-infrared spectroscopy (fNIRS) are also explored for SSR. EEG-based BCI has been studied. Suppes et al. [9] aimed to recognize words from EEG. Signal processing improvements like independent component analysis (ICA) and common spatial pattern (CSP) have enhanced accuracy [10]. Machine learning models are used. Panachakel et al. [11] proposed a deep-learning architecture for decoding imagined speech. fNIRS-based BCI, with better spatial resolution and less electrical noise, measures cerebral blood flow changes [12].

Wearable BCI systems like EEG headsets are a major SSI development. They enable non-invasive, portable brain activity recording for real-time silent speech recognition. Used in medical applications for speech-impaired patients [13, 14], the AlterEgo system shows potential for discreet communication and broader applications like smart home control and virtual reality (VR) interaction.

This chapter will overview the Silent Speech Interface, its core technologies, BCI applications, challenges, and future prospects.

2. Silent speech Interface

Silent Speech Interface (SSI) represents a revolutionary technological paradigm that endeavors to bridge the communication gap by deciphering and reconstructing speech from non-acoustic signals that are intrinsically intertwined with the speech production process. These non-acoustic signals, which can be physiological or neural in origin, offer an alternative conduit for communication, especially in scenarios where audible speech is either unfeasible or undesirable. The overarching goal of SSI is to provide a seamless means of communication that transcends the limitations of traditional audible speech recognition systems. SSI can be conceptually segmented into two principal components: Silent Speech Recognition (SSR), which is primarily concerned with the conversion of non-acoustic signals into textual representations, and Voice Reconstruction, which focuses on the direct synthesis of audible speech from these non-acoustic signals.

A comprehensive description of SSI can be seen in **Figure 1**. Unlike traditional voice recognition systems, which are susceptible to environmental interference and require clear sound signals, SSI technology overcomes these limitations by directly processing physiological and neural signals associated with speech. This innovative approach provides a pathway to more inclusive and adaptive applications in various domains, such as enhancing human-computer interaction, enabling effective communication in loud environments, and supporting assistive communication for those with speech impairments [3].

2.1 Silent speech recognition

Silent Speech Recognition (SSR) is a complex process that involves the translation of non-acoustic signals, generated during the act of speech production but not audible as sound waves, into a textual format. This technology holds the potential to revolutionize communication in numerous scenarios, such as in noisy environments where audible speech may be drowned out, in situations requiring privacy where speaking aloud is not an option, or for individuals with speech impairments who are unable to produce intelligible audible speech. A common SSR processing is outlined in **Figure 2** [16].

In SSR, Electromyography (EMG)-based recognition is commonly used. EMG measures muscle electrical activity involved in speech production. Surface EMG electrodes are placed on relevant skin areas like facial and neck regions to capture and analyze signals from muscle contractions during speech, providing insights into muscle activity patterns for different speech sounds.

For example, Meltzner et al. [6] developed an advanced system with EMG sensors to record facial and neck muscle activity. Their research showed the technical feasibility of using EMG signals for speech recognition and achieved an 8.9% word error rate for continuous phrases with a large vocabulary, highlighting EMG-based SSR's potential. Deng et al. [7] explored combining EMG and acoustic signals for disordered speech recognition. Jou et al. [17] investigated surface electromyography for continuous speech recognition. Walliczek et al. [18] focused on subword unit-based non-audible speech recognition using surface electromyography. Wand and Schultz [19] analyzed phone confusion in EMG-based speech recognition. Wand et al. [20]

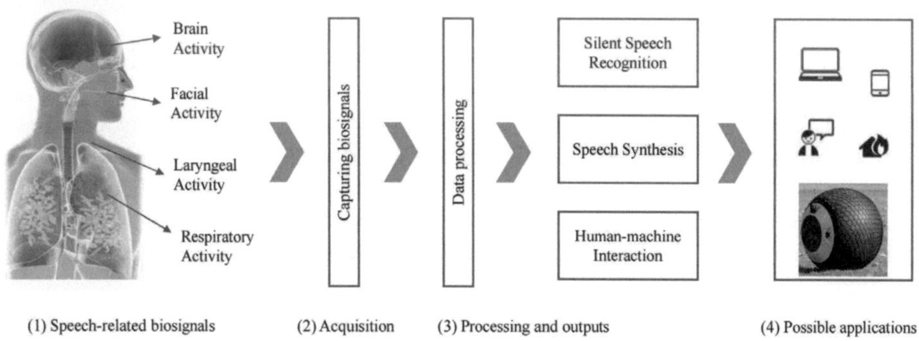

Figure 1.
Biosignal-based spoken communication resulting from (1) speech-related activities of the human body, (2) signal acquisition using various activity-dependent sensor technologies, and (3) biosignal processing including feature extraction followed by output generation for (4) various target applications [15].

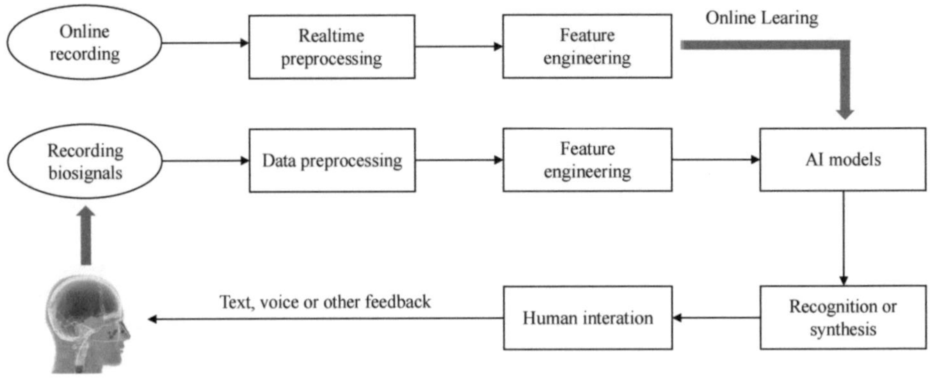

Figure 2.
Common silent speech recognition processing. In this part, neural signals are processed and predicted as texts, which can be further transmitted into voice or video.

explored handling speaking mode varieties in EMG-based speech recognition. Maier-Hein et al. [21] conducted research on session-independent non-audible speech recognition using surface electromyography.

Another key approach in SSR is based on Brain-Computer Interface (BCI) technologies, with Electroencephalography (EEG) being widely studied. EEG records brain electrical activity related to speech production's cognitive and motor processes. Researchers use various signal processing and classification techniques to decode neural patterns associated with speech.

Fitriah et al. [10] surveyed EEG's application in SSR, discussing challenges and advancements. They described how studies use EEG features like frequency bands, event-related potentials, and spatial patterns with machine learning algorithms to enhance recognition accuracy. For example, some focus on specific frequency bands like alpha and beta for discriminative information. Common spatial pattern (CSP) analysis is used to extract features from EEG signals. Suppes et al. [9] were early pioneers in recognizing words from EEG signals. Panachakel et al. [11] proposed a deep-learning architecture for decoding imagined speech from EEG signals with good accuracy results. Wang et al. [22] explored decoding English alphabet letters using EEG phase information. Anumanchipalli et al. [23] made progress in speech synthesis from neural decoding of spoken sentences, relevant to SSR.

Besides EMG and EEG, other methods are investigated in SSR research. Mechanomyography (MMG) detects mechanical vibrations during speech production, offering a complementary view. Imaging techniques like ultrasound and Electromagnetic Articulography (EMA) capture speech articulator movements. Studies have examined combining multiple sensing modalities, like integrating EMG with EEG or MMG, to leverage their strengths and overcome individual limitations, showing promise in improving speech recognition accuracy and reliability in real-world conditions.

2.2 Voice reconstruction

Voice Reconstruction is the process of directly synthesizing audible speech from non-acoustic signals. This technology holds great promise for restoring the natural voice of individuals with speech impairments or for applications that require a more seamless and immediate form of communication.

In voice reconstruction, one key method is mapping articulatory movements and acoustic features. With Electromagnetic Articulography (EMA) or Permanent Magnetic Articulography (PMA) data on speech articulator movement, techniques to estimate acoustic parameters have been developed. Gonzalez et al. [8] proposed a direct speech reconstruction method from articulatory sensor data, using machine learning to model the mapping and getting good results in speech quality and intelligibility. Jorgensen et al. [4] explored articulatory data in subvocal speech recognition related to voice reconstruction.

Another approach uses neural networks to generate speech from non-acoustic signals. Deep-learning models like recurrent neural networks (RNNs) and convolutional neural networks (CNNs) are trained on parallel datasets of non-acoustic and speech recordings to learn the relationships. Janke and Diener [24] converted facial myoelectric signals into speech with deep neural networks, showing real-time generation ability. Diener and Schultz [8] improved this conversion, enhancing speech quality. Anumanchipalli et al. [23] made progress in speech synthesis from neural decoding applicable to voice reconstruction.

Beyond these, efforts to enhance reconstructed voice naturalness and quality exist. Pitch prediction and prosody modeling capture natural speech intonation and rhythm. Speaker adaptation algorithms personalize the reconstructed voice to match user characteristics like speaking style and voice quality. Toth et al. [25] explored synthesizing speech from electromyography with voice transformation techniques for modifying reconstructed voice features.

3. Sensing development

In the domain of Silent Speech Interface (SSI), several types of biosignals play a crucial role in capturing and decoding speech-related information. These biosignals can be broadly categorized into brain activity signals, muscular activity signals, and other related signals. Understanding and effectively sensing these signals are essential for the development and improvement of SSI systems. Typical sensing technologies are EMG, EEG, and electrocorticography (ECoG), which are very prevalent in SSI, as shown in **Figure 3**.

3.1 Brain activity

Brain activity signals matter in SSI, revealing neural speech production processes.

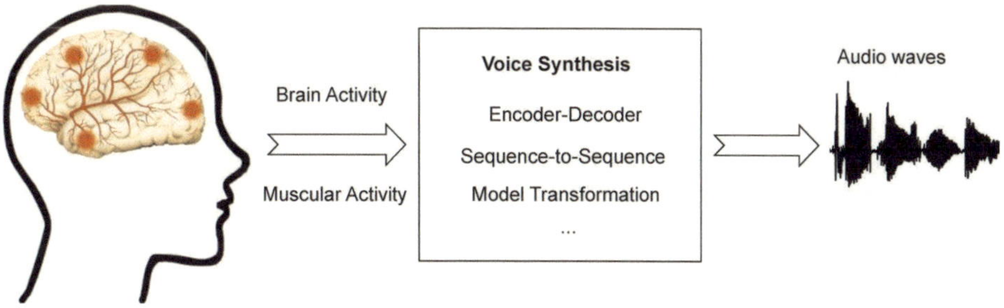

Figure 3.
Voice Reconstruction procedures. In this part, deep neural networks are more often used to generate realistic audio waves than traditional machine learning.

EEG, non-invasive, measures brain electrical activity via scalp electrodes. It's cheap, portable, and easy. Research advanced its speech recognition accuracy. ICA and CSP extract features. Fitriah et al. [10] surveyed EEG in SSR. Machine learning classifies EEG speech patterns. Wang et al. [22] decoded letters from EEG phase. But EEG has low spatial resolution and is prone to artifacts. Still, it suits non-invasive, portable needs like preliminary research or long-term, less precise monitoring.

ECoG is invasive, electrodes on brain surface. It has higher spatial resolution than EEG. Brumberg et al. [2] explored it in BCI speech. Recent studies optimized it. But its invasiveness limits to clinical settings during neurosurgery (e.g., epilepsy treatment), with risks like infection and brain tissue damage.

fNIRS, non-invasive, measures cerebral blood flow and oxygenation. It's a spatial resolution compromise between EEG and ECoG. Gonzalez-Lopez et al. [12] discussed its SSI role. It tolerates movement artifacts better than EEG but has slower temporal resolution. Studies combined it with EEG to understand speech neural correlates and boost recognition accuracy.

Magnetoencephalography (MEG), non-invasive, measures neural magnetic fields. It has higher temporal resolution than EEG and gives complementary neural source information. Less used in SSI due to cost and complexity, but it helps to understand speech brain mechanisms. Some studies used it for speech planning and execution of neural dynamics.

3.2 Muscular activity

Muscular activity signals are another important source of information for SSI. The electrical activity of the muscles involved in speech production can be captured and analyzed to infer the intended speech [15].

Surface electromyography (sEMG) is the most commonly used technique for measuring muscular activity in SSI. It records the electrical signals generated by the muscle fibers when they contract. sEMG electrodes are placed on the skin surface over the relevant muscles, such as those in the face, neck, and throat. In recent years, there have been significant advancements in sEMG technology and its application in speech recognition. Meltzner et al. (2018) [6] developed a system that used sEMG sensors to achieve high accuracy in silent speech recognition. The technique has been refined to improve the signal quality and reduce interference from other muscle activities. However, sEMG signals can be affected by factors such as electrode placement, muscle fatigue, and individual differences in muscle anatomy. Despite these challenges, sEMG is a promising technique for SSI, especially in applications where a non-invasive and relatively simple solution is desired (**Figure 4**).

Intramuscular electromyography (iEMG) involves inserting electrodes directly into the muscle tissue. This technique provides a more accurate measurement of the muscle activity compared to sEMG, as it is less affected by the intervening layers of skin and fat. However, iEMG is more invasive and requires more expertise to perform. It is mainly used in research settings where a more detailed understanding of the muscle activity is required. For example, in studies aiming to understand the fine-grained muscle activation patterns during different speech sounds or in cases where very precise muscle activity measurement is essential for developing more accurate speech recognition models.

Mechanomyography (MMG) is another technique that measures the mechanical vibrations of the muscles during contraction. It can provide complementary

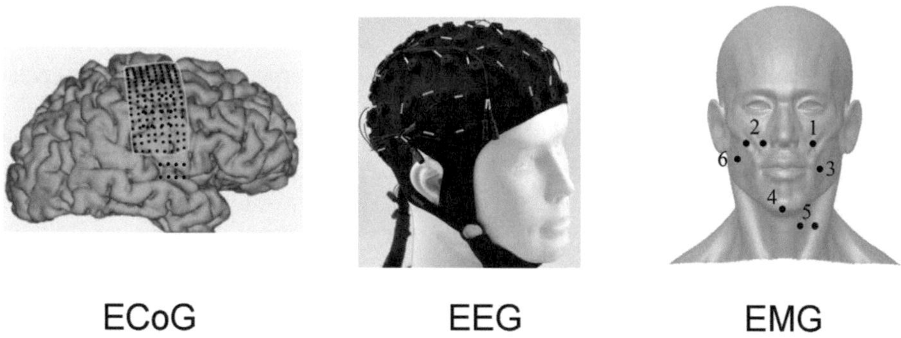

Figure 4.
Typical sensing approaches that can record neural signals for SSI. (1) ECoG (left), which is more accurate but always requires complicated surgery to insert electrodes. (2) EEG (middle) can be considered as the non-invasive method of ECoG. (2) EMG (right), which is easy to acquire, especially for surface EMG. However, the signal is always accompanied by many noises. The black points are recording sites for sEMG.

information to sEMG and iEMG. MMG has been studied in combination with other sensing modalities to improve the performance of SSI systems. For example, studies have explored the integration of MMG with sEMG to enhance the recognition of speech-related muscle activities. The combination of these two modalities can potentially overcome the limitations of each individual method. For instance, when sEMG signals are affected by noise or artifacts, MMG can provide additional information to improve the overall accuracy of speech recognition.

In addition to the above, there is also research on using piezoelectric sensors to measure the mechanical strain in the muscles during speech production. These sensors can provide an alternative way to capture muscular activity and may offer some advantages in terms of sensitivity and durability compared to traditional sEMG electrodes. However, more research is needed to fully understand their potential and optimize their performance in the context of SSI.

3.3 Other sensing technologies

Beyond brain and muscle activity sensing, other technologies can aid SSI research.

Ultrasound imaging visualizes tongue and lip movements [3, 7]. Tracking articulators infers speech sounds. It directly shows production but needs special gear, is complex, and may lack real-time processing for all uses. E.g., in real-time communications, its speed may not give instant feedback.

Optical imaging like near-infrared spectroscopy (NIRS) and optical coherence tomography (OCT) gives speech area tissue and blood flow information. It can monitor speech-related physiological changes but is experimental in SSI. NIRS was probed for muscle/tissue oxygenation in speech, yet more works are needed for reliable speech correlations.

Wearable sensors that capture multiple signals, like sEMG and accelerometer data, help understand speech production. They're handy and unobtrusive for real-world data collection. But sensor integration, data fusion, and power use are issues. Some research devise algorithms to combine sensor data and boost speech recognition accuracy. New materials and designs also cut size and power use while keeping performance.

4. Processing neural signals for SSI

SSI relies on a series of complex algorithms from preprocessing, SSR, to speech reconstruction. In the preprocessing stage, through the use of various filtering methods and effective silence elimination techniques, a high-quality data foundation is provided for subsequent processing. The SSR part employs diverse methods based on different signal sources, including recognition using grammar or phoneme models, in-depth application of neural network technology, conformal learning, etc. In voice reconstruction, based on the neural potential signals of vocalization, a complex model training and processing flow is carried out, and combined with neural vocoders, speech reconstruction is successfully achieved (**Table 1**) [15, 28].

4.1 Feature preprocessing

Feature preprocessing is a crucial step in processing neural signals for SSI, as it helps to enhance the quality of the signals and extract relevant information. This process typically involves several substeps, including preprocessing, signal denoising, and feature extraction. **Table 1** summarized several popular methods used in SSI.

4.1.1 Preprocessing

Preprocessing of neural signals aims to prepare the raw data for further analysis. This may involve tasks such as filtering to remove unwanted frequency components. High-pass filtering can be used to eliminate low-frequency drifts or baseline wander in the signals. Low-pass filtering, on the other hand, helps to remove high-frequency noise that may be present. For example, a low-pass filter can be applied to eliminate electrical interference or muscle artifacts with frequencies above a certain threshold. In some studies, a band-pass filter is also employed to focus on a specific frequency range of interest. Different frequency bands, such as the theta (4–8 Hz), alpha (8–13 Hz), beta (13–30 Hz), or gamma (30–100 Hz) bands, are known to be related to different cognitive and neural processes underlying speech production [3, 10].

Another important preprocessing step is resampling the data to a uniform sampling rate. This ensures that the signals are in a consistent format for subsequent analysis. For instance, if the original signals are recorded at different sampling rates, resampling them to a common rate, say 1000 Hz, simplifies the processing and

Research Area	Method Name	Related References
SSR	SVM	[19]
SSR	HMM	[26]
SSR	CNN	[17]
SSR	LSTM	[11]
SSR	Conformal Prediction	[27]
Voice Reconstruction	Deep Neural Networks	[23]
Voice Reconstruction	Voice Transformation	[25]
Voice Reconstruction	Mapping Acoustic Features	[8]

Table 1.
Some typical SSR and voice reconstruction methods.

comparison of the data. Additionally, detrending the data can be useful to remove any slow-varying trends that might be present. This can be achieved by fitting a polynomial function to the data and subtracting it from the original signal.

4.1.2 Signal denoising

Signal denoising is essential to improve the signal-to-noise ratio and enhance the quality of the neural signals. One commonly used method is wavelet denoising. The principle behind wavelet denoising is to decompose the signal into different frequency components using wavelet transforms. The wavelet transform allows the separation of the signal into approximation and detail coefficients at different scales. By setting a threshold for the detail coefficients, the noise components can be removed while preserving the important signal features. The formula for the wavelet transform of a discrete signal $x[n]$ is given by:

$$W_x(j,k) = \sum_{n=0}^{L-1} x[n]\psi_{j,k}[n] \tag{1}$$

where j represents the scale, k is the translation parameter, $\psi_{j,k}[n]$ is the wavelet function, and L is the length of the signal. After the wavelet decomposition, the thresholding operation can be applied to the detail coefficients. For example, a soft thresholding rule can be used, where the thresholded coefficient $\hat{d}_{j,k}$ is given by:

$$\hat{d}_{j,k} = \begin{cases} sgn\left(d_{j,k}\right)\left(d_{j,k}|-\lambda|\right) & \text{if}|d_{j,k}|>\lambda \\ 0 & \text{if}|d_{j,k}|\leq\lambda \end{cases} \tag{2}$$

where $d_{j,k}$ is the original detail coefficient and λ is the threshold value. Studies such as that by Panachakel et al. [11] have used wavelet-based features for decoding imagined speech from EEG signals and may have incorporated wavelet denoising techniques to improve the quality of the input signals.

Another approach for signal denoising is independent component analysis (ICA). ICA assumes that the observed signals are a linear mixture of independent source signals. The goal is to find the unmixing matrix that can separate the original independent components. The basic ICA model can be expressed as:

$$x = As \tag{3}$$

where x is the observed signal vector, A is the mixing matrix, and s is the vector of independent source signals. ICA algorithms, such as FastICA, are used to estimate the unmixing matrix and recover the independent components. Fitriah et al. [10] discussed the application of ICA in EEG-based SSI for separating the EEG signals from other interfering signals, such as eye movement artifacts and muscle activity.

In addition to wavelet denoising and ICA, high-pass and low-pass filtering can also be used for denoising. High-pass filtering with an appropriate cutoff frequency can remove slow drifts and DC offsets in the signal. Low-pass filtering can attenuate high-frequency noise. A notch filter can be applied to remove specific frequencies, such as the power line frequency (50 Hz or 60 Hz) and its harmonics, which often cause interference in the recorded signals.

4.1.3 Feature extraction

Feature extraction is the process of deriving meaningful features from the preprocessed and denoised signals. In the context of SSI, various features have been proposed. The power spectral density (PSD) provides information about the distribution of signal power across different frequencies. It can be calculated using the Fourier transform. The formula for the PSD of a discrete signal $x[n]$ is:

$$P_{xx}(f) = \frac{1}{N}|X(f)|^2 \quad (4)$$

where $X(f)$ is the Fourier transform of $x[n]$ and N is the length of the signal. The PSD can be used to identify the dominant frequency bands in the neural signals related to speech production. For example, changes in the PSD of the EEG signals in specific frequency bands may indicate different speech-related neural activities.

Another important feature is the autoregressive (AR) model coefficients. The AR model assumes that the current value of a signal can be predicted based on its previous values. The AR model of order p for a signal $x[n]$ is given by:

$$x[n] = \sum_{i=1}^{p} a_i x[n-i] + e[n] \quad (5)$$

where a_i are the AR coefficients and $e[n]$ is the error term. The AR coefficients can capture the temporal dynamics of the signal and have been used in some SSI studies to represent the characteristics of the neural signals related to speech. For instance, in the analysis of EEG signals for speech recognition, the AR coefficients can provide information about the rhythmic patterns of the neural activity.

Common spatial pattern (CSP) is a widely used method in EEG signal processing for feature extraction, especially in the context of brain-computer interfaces. CSP aims to find spatial filters that maximize the difference in variance between two classes (e.g., different speech tasks or conditions). The algorithm calculates a transformation matrix W that projects the multichannel EEG data X into a new space where the variance of the signals related to one class is maximized and the variance of the signals related to the other class is minimized. The projected data $Z = WX$ can then be used to extract features, such as the logarithm of the variance of the projected signals. The formula for calculating the CSP spatial filters involves solving an eigenvalue problem. Let C_1 and C_2 be the covariance matrices of the two classes of EEG data. The combined covariance matrix $C = C_1 + C_2$ is first calculated. Then, the eigenvalues λ and eigenvectors v of $C^{-1}C_1$ are computed. The spatial filters W are then constructed using the eigenvectors corresponding to the largest and smallest eigenvalues.

4.2 Silent speech recognition

Silent Speech Recognition (SSR) involves the development and application of algorithms to decode the silent speech from the neural signals. The following are the typical methods used in SSR with detailed descriptions and formulas. Here, we introduced several advanced methods for SSR.

4.2.1 Convolutional neural network (CNN)

CNNs can automatically extract hierarchical features from the neural signals. The architecture of a CNN typically consists of convolutional layers, pooling layers, and fully connected layers.

In the convolutional layer, filters (also called kernels) are applied to the input signals to extract local features. Let x_{ij} be an element of the input signal (e.g., a 2D (two-dimensional) representation of the EEG data over time and channels) and w_{mn} be an element of the filter. The output of a convolutional layer y_{ij} at a particular location (i, j) is calculated as:

$$y_{ij} = f\left(\sum_{m=0}^{M-1}\sum_{n=0}^{N-1} w_{mn}x_{i+m,j+n} + b\right) \tag{6}$$

where M and N are the dimensions of the filter, b is the bias term, and f is a non-linear activation function, such as the rectified linear unit (ReLU) $f(z) = \max(0, z)$.

Pooling layers are used to downsample the feature maps. For example, in max-pooling, the maximum value within a local neighborhood is selected as the output. This helps in reducing the computational complexity and also provides some degree of translation invariance.

The fully connected layers are used for classification. The output of the last convolutional or pooling layer is flattened and connected to the fully connected layers. The weights and biases in the fully connected layers are adjusted during the training process to minimize a loss function, such as the cross-entropy loss for classification tasks. CNNs have been used in SSR to learn complex spatial and temporal patterns in the neural signals related to speech. For example, in processing EEG data, the convolutional layers can capture the local patterns in the signals across different channels and time steps, and the fully connected layers can then classify these patterns into different speech categories [17].

4.2.2 Recurrent neural network (RNN)-long short-term memory (LSTM)

RNNs, such as LSTM networks, are suitable for processing sequential data and can capture the long-term dependencies in the neural signals. The LSTM cell has specific update equations for gates (input, forget, output) and cell state to handle sequential information effectively.

The input gate i_t, forget gate f_t, output gate o_t, and cell input activation g_t are calculated as follows:

$$\begin{aligned} i_t &= \sigma(W_i x_t + U_i h_{t-1} + b_i) \\ f_t &= \sigma(W_f x_t + U_f h_{t-1} + b_f) \\ o_t &= \sigma(W_o x_t + U_o h_{t-1} + b_o) \\ g_t &= \tanh(W_g x_t + U_g h_{t-1} + b_g) \end{aligned} \tag{7}$$

where x_t is the input at time t, h_{t-1} is the hidden state at the previous time step, W and U are weight matrices, b are bias vectors, and σ is the sigmoid function. The cell state c_t is updated as:

$$c_t = f_t c_{t-1} + i_t g_t \tag{8}$$

And the hidden state h_t is calculated as:

$$h_t = o_t \tanh(c_t) \tag{9}$$

LSTMs have been applied in SSR to handle the sequential nature of the neural signals during speech production. For example, in processing the time series of EEG or other neural signals, LSTMs can remember the context from previous time steps and use it to predict the current and future speech-related patterns [11].

4.2.3 Conformal prediction

Conformal Prediction is a method used for uncertainty quantification in machine learning. It can be combined with various base models, such as neural networks, to provide more reliable predictions.

In the context of SSR, the main idea of Conformal Prediction is to calculate a nonconformity score for each data point. Let $Z = \{(x_1,y_1), (x_2,y_2), \cdots, (x_n,y_n)\}$ be the training dataset, where x_i is the feature vector and y_i is the corresponding label. For a new data point x_{new}, we first make a prediction y_{pred} using the base model. Then, we calculate the nonconformity score a_i for each training data point (x_i,y_i) and the new data point (x_{new},y_{pred}) using a nonconformity measure.

One common nonconformity measure is the difference between the predicted value and the actual value. For example, if the base model predicts a continuous value, we can use $a_i = |y_i - \hat{y}_i|$, where $\hat{y}_i$ is the predicted value for x_i. If the problem is a classification problem, we can use the probability of the predicted class or a distance measure in the feature space.

After calculating the nonconformity scores, we sort them in ascending order. Let $a_{(1)} \leq a_{(2)} \leq \cdots \leq a_{(n)}$ be the sorted nonconformity scores. We then find the index k such that $\frac{k}{n+1} \approx \epsilon$, where ϵ is the significance level (a value between 0 and 1). The prediction set for the new data point x_{new} is then all the labels y such that the nonconformity score of (x_{new},y) is less than or equal to $a_{(k)}$.

In other words, the prediction set is $\{y : a(x_{new},y) \leq a_{(k)}\}$. If the prediction set contains only one label, then we have a unique prediction. If it contains multiple labels, then we have a set of possible predictions, and the uncertainty of the prediction is reflected by the size of the set.

The advantage of Conformal Prediction is that it provides a way to quantify the uncertainty of the prediction and can give more reliable results, especially in situations where the data are noisy or the model is not very accurate. It has been applied in various fields, including speech recognition, to improve the robustness of the predictions [27].

4.3 Voice synthesis using neural signals

Voice synthesis from neural signals aims to reconstruct audible speech from the neural activity patterns. The following are the methods related to voice synthesis and reconstruction.

4.3.1 Direct conversion using deep neural networks

This method involves training a deep neural network to directly map the neural signals (e.g., facial myoelectric signals) to speech waveforms. Let x be the input neural signal vector and y be the corresponding speech waveform. The neural network is trained to minimize a loss function that measures the difference between the predicted speech waveform $\hat{y}$ and the actual speech waveform y. One commonly used loss function is the mean squared error (MSE):

$$MSE = \frac{1}{N}\sum_{i=1}^{N} (y_i - \hat{y}_i)^2 \tag{10}$$

where N is the number of samples in the speech waveform. The neural network architecture can consist of multiple layers, such as fully connected layers or a combination of convolutional and recurrent layers. The weights and biases in the network are adjusted during training to minimize the MSE. For example, Diener and Schultz [8] used deep neural networks to achieve this conversion, where the input was the facial myoelectric signals and the output was the synthesized speech. The network learned the complex mapping between the myoelectric signals and the speech characteristics, such as the fundamental frequency, formants, and amplitudes.

Suppose we have a simple neural network with one hidden layer for this direct conversion. Let the input layer have n neurons corresponding to the features of the neural signal, the hidden layer have m neurons, and the output layer have p neurons corresponding to the samples of the speech waveform. The weights between the input and hidden layers are denoted as W_1 (with dimensions $m \times n$), and the biases as b_1 (a vector of length m). The weights between the hidden and output layers are W_2 (with dimensions $p \times m$), and the biases as b_2 (a vector of length p).

The output of the hidden layer h is calculated as:

$$h = f(W_1 x + b_1) \tag{11}$$

where f is a non-linear activation function, such as the sigmoid function $f(z) = \frac{1}{1+e^{-z}}$ or the ReLU function $f(z) = \max(0, z)$.

The output of the network $\hat{y}$ is then:

$$\hat{y} = W_2 h + b_2 \tag{12}$$

During training, the weights and biases are updated using an optimization algorithm such as gradient descent. The gradients of the loss function with respect to the weights and biases are calculated, and then the weights and biases are updated in the opposite direction of the gradients to minimize the loss.

4.3.2 Speech synthesis from neural decoding

This approach involves decoding the neural activity related to speech production and using it to generate speech. Anumanchipalli et al. [23] achieved significant progress in this area. The method typically involves training a model to learn the relationship between the neural activity patterns and the corresponding speech features. Let z be the neural activity vector and s be the speech feature vector (e.g., acoustic parameters like F0, formants). The model is trained to minimize a loss function that measures the difference between the predicted speech features $\hat{s}$ and the actual speech features s.

For example, if we are modeling the fundamental frequency $F0$, and let s_i be the actual $F0$ value at a particular time step i and $\hat{s}_i$ be the predicted value. A common loss function for this could be the mean absolute error (MAE):

$$MAE = \frac{1}{T}\sum_{i=1}^{T} |s_i - \hat{s}_i| \tag{13}$$

where T is the total number of time steps.

The model can be a neural network or other machine learning models. Once the model is trained, it can generate speech features from the neural activity patterns. These speech features can then be used to synthesize speech using techniques such as vocoding or waveform synthesis methods.

If we consider a neural network model for this neural decoding, similar to the architecture described above for direct conversion, the input is the neural activity vector z and the output is the speech feature vector s. The hidden layers can capture the complex relationships between the neural activity and the speech features.

4.3.3 Synthesizing speech from electromyography using voice transformation techniques

Toth et al. [25] explored this method. The process involves transforming the electromyography (EMG) signals into speech using specific voice transformation algorithms.

First, the EMG signals are preprocessed and relevant features are extracted. Let x_{EMG} be the EMG signal. After preprocessing, we get a feature vector f_{EMG}. This feature extraction could involve methods, such as calculating the amplitude, frequency content, or other characteristics of the EMG signal.

Then, a transformation model T is trained to map these EMG features to acoustic parameters of speech, such as the fundamental frequency (F0), formants, and amplitudes. The predicted acoustic parameters $\hat{p}$ are given by:

$$\hat{p} = T(f_{EMG}) \tag{14}$$

For example, if T is a linear regression model, and f_{EMG} has n features, and we are predicting the fundamental frequency $F0$ (a single value), the model can be represented as:

$$\hat{F0} = w_0 + w_1 f_{EMG1} + w_2 f_{EMG2} + \cdots + w_n f_{EMGn} \tag{15}$$

where w_i are the regression coefficients.

Once the acoustic parameters are predicted, they are used to synthesize the speech waveform. This can be done using techniques like vocoding. For example, in a simple sinusoidal model vocoder, the speech waveform y at a particular time t can be synthesized as:

$$y(t) = \sum_{k=1}^{K} A_k(t)\sin\left(2\pi f_k(t)t + \varphi_k(t)\right) \tag{16}$$

where K is the number of harmonics, $A_k(t)$ is the amplitude of the kth harmonic, $f_k(t)$ is the frequency of the kth harmonic (which can be related to the predicted F0 and formants), and $\varphi_k(t)$ is the phase of the kth harmonic.

4.3.4 Mapping acoustic features for speech reconstruction

Gonzalez et al. [8] proposed methods for directly reconstructing speech from articulatory sensor data. The articulatory sensor data, which could be obtained from techniques like Electromagnetic Articulography (EMA) or Permanent Magnetic Articulography (PMA), capture the movement of the speech articulators.

Let a be the vector of articulatory sensor data and s be the vector of acoustic parameters. A mapping model M is trained to predict the corresponding acoustic parameters from the articulatory sensor data. The predicted acoustic parameters $\hat{s}$ are given by:

$$\hat{s} = M(a) \tag{17}$$

If M is a neural network, similar to the previous architectures, it can have multiple layers to learn the complex mapping. The loss function used during training can be the mean squared error between the predicted and actual acoustic parameters, similar to the direct conversion method described above.

$$MSE = \frac{1}{N}\sum_{i=1}^{N}(s_i - \hat{s}_i)^2 \tag{18}$$

The trained model can then be used to reconstruct speech from new articulatory sensor data by predicting the acoustic parameters and synthesizing the speech waveform using appropriate techniques, such as the vocoding methods described earlier.

5. Use cases

Silent Speech Interface (SSI) holds great promise with a multitude of potential application scenarios, as explored in the research of Biosignal-Based Spoken Communication. These applications span across various domains, aiming to address different communication needs and challenges.

5.1 Restoration of speech disability

In speech disability restoration, SSI has great potential. For laryngectomy patients, Meltzner et al. [6] used sEMG sensors on facial/neck muscles, achieving 8.9% word error rate for large vocabulary continuous phrases. For those with neurological disorders like amyotrophic lateral sclerosis (ALS), BCI-based SSI like Brumberg et al. [2] ECoG study can decode neural speech intentions. Children with congenital speech disorders can benefit. Smith et al. [29] combined sEMG and EEG to develop a system for non-verbal kids.

For stroke survivors with aphasia, Musso et al. [30] used an SSI system with machine learning in speech therapy, enhancing speech production and comprehension.SSI performance must be high. Accuracy is key. Current research focuses on better signal processing, personalized models, and more user-friendly, less invasive systems.

5.2 Enhancement for noisy speech

In noisy settings, traditional speech recognition fails. SSI helps. In industry, Jou et al. [17] used sEMG for speech recognition, achieving better accuracy than audio methods as sEMG resists noise.

In military, silent speech SSI systems with sensors like sEMG and accelerometers are developed. Brown et al. [31, 32] showed their potential in a simulated military operation scenario, enhancing stealth and efficiency.

The AlterEgo SSI device aids factory workers. It uses electrode technology to capture face/neck signals, transmitting recognized speech wirelessly. Field trials prove it boosts efficiency and cuts miscommunication risks.

For noisy speech enhancement, SSI needs noise resistance, high sensitivity, and quick response. Current research focuses on signal filtering, feature extraction, multimodal sensing, and advanced ML for classification.

5.3 Silent speech in private or military situations

In private scenarios like libraries or quiet offices, SSI enables silent communication. A research team's prototype [12, 13, 16] detects mouth and jaw movements, translating them into text or speech. User trials showed high accuracy in simple conversations, with the system well-regarded for quiet public space use. In courtrooms, lawyers can use SSI with a collar sensor for discreet client communication.

In military and security applications, silent speech is vital for covert operations. SSI systems with advanced encryption and secure transmission are in development. For these uses, SSI must be secure, private, and reliable. Current research focuses on integrating encryption and protocols, while enhancing portability and ease of use for diverse civilian and military scenarios.

6. Challenges and future directions

Silent Speech Interface (SSI) technology, despite its significant potential, is still in the developmental stage and faces a multitude of challenges across various aspects, ranging from fundamental research to practical applications, and from algorithmic models to the deployment of actual devices [33]. These challenges not only impede the current progress of SSI but also define the key areas for future research and improvement. Overcoming these hurdles is crucial for SSI to realize its full potential and gain wider acceptance and application in different fields.

6.1 Bio-sensing noises

Bio-sensing noises pose a substantial obstacle to the accurate acquisition and interpretation of neural and muscular signals for SSI [12, 15]. In real-world scenarios, the signals captured by sensors are often contaminated with various types of noises. For instance, during the measurement of electroencephalography (EEG) signals, environmental electromagnetic interference can introduce artifacts that mimic or distort the actual neural activity related to speech. Similarly, surface electromyography (sEMG) signals used to detect muscle activity can be affected by motion artifacts, electrode impedance changes, and electrical noise from surrounding equipment.

The presence of these noises can significantly degrade the quality of the input data for SSI systems, leading to inaccurate feature extraction and subsequent misclassification or incorrect speech reconstruction. To address this issue, advanced signal processing techniques are required. One approach is to develop more sophisticated noise cancelation algorithms that can adaptively filter out the unwanted noise components while preserving the integrity of the relevant biosignals. For example, wavelet-based denoising methods can be further enhanced to better handle the non-stationary and time-varying nature of bio-sensing noises [11]. Additionally, improving the design and placement of sensors can help reduce the susceptibility to noise. For example, using more advanced electrode materials and configurations in EEG recordings or optimizing the attachment of sEMG electrodes to minimize motion artifacts [17].

6.2 Model generalization

Model generalization is a critical challenge in the development of SSI systems. Current machine learning and deep-learning models often struggle to generalize well across different individuals, speaking conditions, and environments. This is because the characteristics of the biosignals used for SSI can vary significantly from person to person due to differences in anatomy, physiology, and speaking habits.

For example, a model trained on the EEG or sEMG data of a specific group of individuals may perform poorly when applied to a new user with different signal patterns. To enhance model generalization, more diverse and representative datasets are needed for training. These datasets should cover a wide range of ages, genders, ethnicities, and speech disorders to ensure that the models can learn the common and invariant features across different populations. Transfer learning and domain adaptation techniques can also be explored. By leveraging knowledge learned from related tasks or datasets, models can potentially adapt more quickly and accurately to new users or conditions. For instance, a pretrained model on a large general speech dataset can be fine-tuned for a specific SSI application, reducing the amount of training data required from the new user and improving generalization performance [34].

6.3 Evaluation benchmarks, datasets, and standards

The lack of comprehensive and standardized evaluation datasets and benchmarks is a significant hurdle in the progress of SSI research. Currently, different studies often use their own custom datasets, which vary in size, quality, and the types of speech tasks and conditions they represent. This makes it difficult to directly compare the performance of different SSI algorithms and systems objectively.

To promote the development and evaluation of SSI technologies, there is an urgent need to establish large-scale, publicly available datasets that encompass a wide variety of speech scenarios, including different languages, speaking styles, and noise conditions. These datasets should also be accompanied by standardized evaluation metrics and protocols. For example, common measures, such as word error rate, speech reconstruction accuracy, and user satisfaction, should be clearly defined and consistently applied across different studies. Additionally, guidelines for data collection, preprocessing, and annotation should be developed to ensure the reproducibility and comparability of research results. This would enable researchers to more effectively evaluate the performance of their algorithms and identify areas for improvement, ultimately accelerating the development of more reliable and effective SSI systems [31].

6.4 Data security and social ethics

As SSI systems involve the collection and processing of sensitive personal data, such as neural and muscular activity patterns, data security and social ethics are of paramount importance. The potential for unauthorized access to these data raises concerns about privacy invasion and misuse. For example, if an SSI device is hacked, an attacker could gain access to an individual's speech-related neural signals, potentially revealing personal information or even being used for malicious purposes, such as identity theft or impersonation.

To safeguard data security, robust encryption techniques should be employed to protect the transmission and storage of biosignal data. Access controls and authentication mechanisms need to be implemented to ensure that only authorized users and devices can interact with the SSI system. From a social ethics perspective, it is essential to have transparent policies and informed consent procedures in place. Users should be fully aware of how their data will be used, stored, and shared, and have the right to control and manage their personal information. Additionally, ethical considerations regarding the potential impact of SSI on social interactions and human behavior need to be carefully examined. For example, the use of SSI in public spaces may raise questions about the boundaries of privacy and the potential for involuntary communication or surveillance. Addressing these data security and social ethics challenges is crucial for the responsible development and deployment of SSI technologies [29].

7. Conclusions

This comprehensive review explored the Silent Speech Interface (SSI). In sensing development, various biosignals like EEG, ECoG, fNIRS, MEG for brain activity and sEMG, iEMG, MMG, piezoelectric sensors for muscular activity, along with ultrasound and optical imaging, are crucial. Their refinement is key for accurate silent speech capture.

The neural signal processing for SSI has multiple steps. Feature preprocessing (preprocessing, denoising, extraction) is the base. SSR includes algorithms from SVM, HMM to CNN, LSTM, and Conformal Prediction. Voice synthesis has different methods. SSI has wide use cases, from restoring speech ability in various patient groups to enhancing noisy speech in multiple settings and enabling silent communication in private and military scenarios. But it faces challenges like bio-sensing noises, model generalization, lack of evaluation standards, and data security and ethics concerns.

The future of SSI is promising. Advancing sensor technology will lead to better sensing and integration. Powerful models will improve performance. Standardized evaluation frameworks will boost research. With growing awareness of privacy and ethics, SSI will develop responsibly. Long-term, it could revolutionize communication, especially for speech-impaired, and expand human-computer interaction in many fields.

Conflict of interest

The authors declare no conflict of interest.

Author details

Ming Zhang[1,2*†], Yuan Yuan[2†] and Shuo Zhang[3]

1 Huzhou Institute of Zhejiang University, Huzhou, China

2 Shezhen BrainCLOS, Shenzhen, China

3 Lincoln University, Canterbury, New Zealand

*Address all correspondence to: drystan@zju.edu.cn

† These authors contributed equally.

References

[1] Morse MS, O'Brien EM. Research summary of a scheme to ascertain the availability of speech information in the myoelectric signals of neck and head muscles using surface electrodes. Computers in Biology and Medicine. 1986;**16**(6):399-410. DOI: 10.1016/0010-4825(86)90064-8 [Accessed: October 28, 2024]

[2] Brumberg JS, Nieto-Castanon A, Kennedy PR, Guenther FH. Brain-computer interfaces for speech communication. Speech Communication. 2010;**52**(4):367-379. DOI: 10.1016/j.specom.2010.01.001 [Accessed: October 28, 2024]

[3] Hueber T, Chollet G, Denby B, Stone M. Acquisition of ultrasound, video and acoustic speech data for a silent-speech interface application. Proceedings of ISSP. 2008:365-369. Available from: https://api.semanticscholar.org/CorpusID:17377079 [Accessed: October 28, 2024]

[4] Jorgensen C, Lee DD, Agabon S. Sub auditory speech recognition based on EMG signals. Proceedings of the International Joint Conference on Neural Network. 2003;**4**:3128-3133. DOI: 10.1109/IJCNN.2003.1224072 [Accessed: October 28, 2024]

[5] Manabe H. Unvoiced speech recognition using EMG—Mime speech recognition. In: CHI'03 Extended Abstracts on Human Factors in Computing Systems. New York, NY, United States: Association for Computing Machinery; 2003. pp. 794-795. DOI: 10.1145/765891.765996 [Accessed: October 28, 2024]

[6] Meltzner GS, Heaton JT, Deng Y, De Luca G, Roy SH, Kline JC. Silent speech recognition as an alternative communication device for persons with laryngectomy. IEEE/ACM Transactions on Audio, Speech, Language Process. 2018;**25**(12):2386-2398. DOI: 10.1109/TASLP.2017.2740000 [Accessed: October 28, 2024]

[7] Luo J, Wang J, Cheng N, Jiang G, Xiao J. End-to-end silent speech recognition with acoustic sensing. In: 2021 IEEE Spoken Language Technology Workshop (SLT). New York, NY, United States: IEEE; 2021. pp. 606-612. DOI: 10.1109/SLT48900.2021.9383622 [Accessed: October 28, 2024]

[8] Diener L, Schultz T. Direct conversion from facial myoelectric signals to speech using deep neural networks. In: IJCNN. New York, NY, United States: IEEE; 2018. pp. 1-7. DOI: 10.1109/IJCNN.2015.7280404 [Accessed: October 28, 2024]

[9] Suppes P, Lu ZL, Han B. Brain wave recognition of words. Proceedings of the National Academy of Sciences. 1997; **94**(26):14965-14969. DOI: 10.1073/pnas.94.26.14965 [Accessed: October 28, 2024]

[10] Fitriah N, Zakaria H, Rajab TLE. EEG-based silent speech Interface and its challenges: A survey. International Journal of Advanced Computer Science and Applications. 2022;**13**(11):625-635. DOI: 10.14569/IJACSA.2022.0131173 [Accessed: October 28, 2024]

[11] Panachakel JT, Ramakrishnan AG, Ananthapadmanabha TV. Decoding imagined speech using wavelet features and deep neural networks. In: 2019 IEEE 16th India Council International Conference (INDICON), Rajkot, India. New York, NY, United States: IEEE; 2019. pp. 1-4. DOI: 10.1109/INDICON47234.2019.9028925 [Accessed: October 28, 2024]

[12] Gonzalez-Lopez JA, Gomez-Alanis A, Martin Donas JM, Perez-Cordoba JL, Gomez AM. Silent speech interfaces for speech restoration: A review. IEEE Access. 2020;**8**:177995-178021. DOI: 10.1109/ACCESS.2020.3026579 [Accessed: October 28, 2024]

[13] Kapur A, Kapur S, Maes P. AlterEgo: A personalized wearable silent speech Interface. In: IUI 2018, Mar. 7–11, 2018. Tokyo, Japan: ACM; 2018. pp. 1-10. DOI: 10.1145/3172944.3172977 [Accessed: October 28, 2024]

[14] Srivastava T, Winters RM, Gable T, Wang YT, LaScala T, Tashev IJ. Whispering wearables: Multimodal approach to silent speech recognition with head-worn devices. In: Proceedings of the 26th International Conference on Multimodal Interaction. New York, NY, United States: Association for Computing Machinery; 2024. pp. 214-223. DOI: 10.1145/3678957.3685720 [Accessed: October 28, 2024]

[15] Schultz T, Wand M, Hueber T, Krusienski DJ, Herff C, Brumberg JS. Biosignal-based spoken communication: A survey. IEEE/ACM Transactions on Audio, Speech, and Language Processing. 2017;**25**(12):2257-2271. DOI: 10.1109/TASLP.2017.2752365 [Accessed: October 28, 2024]

[16] Zhang M, Wang Y, Zhang W, Yang M, Luo Z, Li G. Inductive conformal prediction for silent speech recognition. Journal of Neural Engineering. 2020;**17**(6):066019. [Accessed: October 28, 2024]. DOI: 10.1088/1741-2552/ab7ba0

[17] Jou SC, Schultz T, Walliczek M, Kraft F, Waibel A. Towards continuous speech recognition using surface electromyography. In: Interspeech 2006 - ICSLP. Pittsburgh, PA, USA: ISCA; 2006. pp. 1-4. DOI:10.21437/interspeech.2006-212

[18] Walliczek M, Jou SH, Liu CC, Jeng SK. Sub-word unit based non-audible speech recognition using surface electromyography. In: 2006 IEEE International Conference on Multimedia and Expo. Pittsburgh, PA, United States: ISCA; 2006. pp. 2173-2176. DOI: 10.21437/Interspeech.2006-425 [Accessed: October 28, 2024]

[19] Wand M, Schultz T. Analysis of phone confusion in EMG-based speech recognition. In: 2011 IEEE International Conference on Acoustics, Speech and Signal Processing (ICASSP). New York, NY, United States: IEEE; 2011. pp. 757-760. DOI: 10.1109/ICASSP.2011.5946514 [Accessed: October 28, 2024]

[20] Wand M, Janke M, Schultz T. Tackling speaking mode varieties in EMG-based speech recognition. IEEE Transactions on Biomedical Engineering. 2014;**61**(10):2515-2526. DOI: 10.1109/TBME.2014.2319000 [Accessed: October 28, 2024]

[21] Maier-Hein L, Meinecke FC, Hueber T. Session independent non-audible speech recognition using surface electromyography. In: 2005 IEEE International Conference on Acoustics, Speech and Signal Processing (ICASSP). New York, America: IEEE; 2005. pp. 1081-1084. DOI: 10.1109/ASRU.2005.1566521 [Accessed: October 28, 2024]

[22] Wang Y, Liu Y, Zhang H, Hu D. Decoding English alphabet letters using EEG phase information. In: 2018 40th Annual International Conference of the IEEE Engineering in Medicine and Biology Society (EMBC). Lausanne, Switzerland: Frontiers Media SA; 2018.

pp. 4779-4782. DOI: 10.3389/fnins.2018.00062 [Accessed: October 28, 2024]

[23] Anumanchipalli GK, Chartier J, Chang EF. Speech synthesis from neural decoding of spoken sentences. Nature. 2019;**568**(7753):493-498. DOI: 10.1038/s41586-019-1119-1 [Accessed: October 28, 2024]

[24] Janke M, Diener L. EMG-to-speech: Direct generation of speech from facial electromyographic signals. IEEE/ACM Transactions on Audio, Speech, and Language Processing. 2017;**25**(12):2375-2385. DOI: 10.1109/TASLP.2017.2738568 [Accessed: October 28, 2024]

[25] Toth AR, Wand M, Schultz T. Synthesizing speech from electromyography using voice transformation techniques. In: 10th Annual Conference of the International Speech Communication Association, INTERSPEECH 2009, Brighton, United Kingdom, September 6–10. Brighton, UK: ISCA; 2009. pp. 652-655. DOI: 10.21437/Interspeech.2009-229 [Accessed: October 28, 2024]

[26] Makeig S, Debener S, Onton J, Delorme A. Mining event-related brain dynamics. Trends in Cognitive Sciences. 2004;**8**(5):204-210. DOI: 10.1016/j.tics.2004.03.008 [Accessed: October 28, 2024]

[27] Wang Y, Zhang M, Wu R, Wang H, Luo Z, Li G. Speech neuromuscular decoding based on spectrogram images using conformal predictors with Bi-LSTM. Neurocomputing. 2021;**451**:25-34. DOI: 10.1016/j.neucom.2021.03.025 [Accessed: October 28, 2024]

[28] Shandiz AH. Improvements of Silent Speech Interface Algorithms. Hungary: Szeged University; 2023 [Accessed: October 28, 2024]

[29] Nissenbaum H. Privacy in Context: Technology, Policy, and the Integrity of Social Life. Stanford, USA: Stanford University Press; 2010. DOI: 10.1080/03623319.2023.2232223 [Accessed: October 28, 2024]

[30] Musso M, Hübner D, Schwarzkopf S, Bernodusson M, LeVan P, Weiller C, et al. Aphasia recovery by language training using a brain–computer interface: A proof-of-concept study. Brain Communications. 2022;**4**(1):fcac008. DOI: 10.1093/braincomms/fcac008 [Accessed: October 28, 2024]

[31] Freitas J, Teixeira A, Dias MS, Silva S, et al. An Introduction to Silent Speech Interfaces. Cham, Switzerland: Springer; 2017

[32] Huang B, Shao Y, Zhang H, Wang P, Chen X, Li Z, et al. Design and implementation of a silent speech recognition system based on sEMG signals: A neural network approach. Biomedical Signal Processing and Control. 2024;**92**:106052. DOI: 10.1016/j.bspc.2024.106052 [Accessed: October 28, 2024]

[33] Denby B, Schultz T, Honda K, Hueber T, Gilbert JM, Brumberg JS. Silent speech interfaces. Speech Communication. 2010;**52**(4):270-287. DOI: 10.1016/j.specom.2009.08.002 [Accessed: October 28, 2024]

[34] Goodfellow I, Bengio Y, Courville A. Deep Learning. Cambridge, Massachusetts, USA: MIT Press; 2016 [Accessed: October 28, 2024]

Section 4

Machine Ethics

Chapter 9

Perspective Chapter: Toward Effective Ethical AI – Educating AI Robots through Machine Ethics Theories

Xi Yue, Paracha Samiullah, Hands Caroline, Ioannis Angelakis and Sun Ruoyu

Abstract

The advancements in Artificial Intelligence (AI) and robotics have profoundly transformed human productivity and daily life. The rapid rise of AI robots has contributed to the growth of humans, yet the lack of intelligence has sparked concerns about related ethical challenges. At present, there is no consensus within the academic community regarding the ethical design of AI robots. The primary aim of this chapter is to explore existing ethical problems and design issues in AI robots. This chapter analyzes the case of the AI robot Xiaoice and investigates the experiences of users when using AI robots. Preliminary results indicate that AI robots have inspired users' understanding of daily life and education, equipping them with the necessary technical skills and significantly improving efficiency in handling mundane tasks. The chapter also examines the challenges and difficulties faced in the implementation of AI, such as algorithmic bias, privacy protection, distrust between humans and AI robots, and other common technological issues. A survey of 97 users revealed that they believe there are still many areas for improvement in the design of existing AI robots, providing insights for the future development and applications in this field.

Keywords: machine ethics, ethical robots, intelligent agents, AI ethics and trustworthiness, human-robot interaction, relationship

1. Introduction

The socio-technical impact of robotics and Artificial Intelligence (AI) is increasingly significant, as evidenced by their integration into a wide range of organizational and industrial processes [1]. Moreover, the growing prominence of AI in everyday life, coupled with the emergence of autonomous intelligent agents within industrial settings, underscores its expanding role [1]. Developing ethical design frameworks and building trust toward AI robots is becoming increasingly critical. AI refers to systems designed to simulate or perform tasks typically associated with human intelligence, including learning, reasoning, understanding language, problem-solving, and image recognition [2, 3]. AI robots are intelligent machines that integrate AI technology and robotics, including virtually and physically. Both AI and robotics are digital technologies (**Figure 1**) set to significantly influence human advancement in the near future [4]. The development of AI and robotics have significantly improved societal productivity while greatly enhancing the convenience and quality of human life. For example, Xiaoyi of Huawei and Siri of Apple use AI-powered natural language processing to help users with daily tasks, such as playing music or managing schedules.

But at the same time, the rapid advancement of AI robots has also posed challenges [5], particularly in areas such as privacy security, social ethics, and human-AI interactions. Key issues include algorithmic bias, quasi-subjectivity dilemmas, technological unemployment, privacy disclosure and distrust between humans and AI robots. Furthermore, public concerns and debates have intensified regarding whether AI robots should be "educated" on moral principles and, if so, how such principles can be effectively implemented [6–8].

This chapter focuses on addressing two critical questions:

RQ1: "*How to educate a sense of morals through machine ethics during the development process of AI robots?*"

RQ2: "*How to effectively address the specific challenge of building emotional connections and trust between users and AI robots?*"

It is worthy mentioned that the conception of mental functioning of AI was naive, so familiar with philosophical approaches to humans is important [9]. The development of machine ethics provides a framework for considering moral rules, which were the rational foundation of morality [10], and ethical responsibilities during the design process of AI robots. This theory explores not only the ethical considerations surrounding machine development but also the integration of moral functions into the practical applications [11], which primarily draw from Kant's deontological ethics and Bentham's utilitarian philosophy (**Figure 2**). Deontology, an ethical theory that emphasizes duty and responsibility, evaluates moral actions based on their adherence

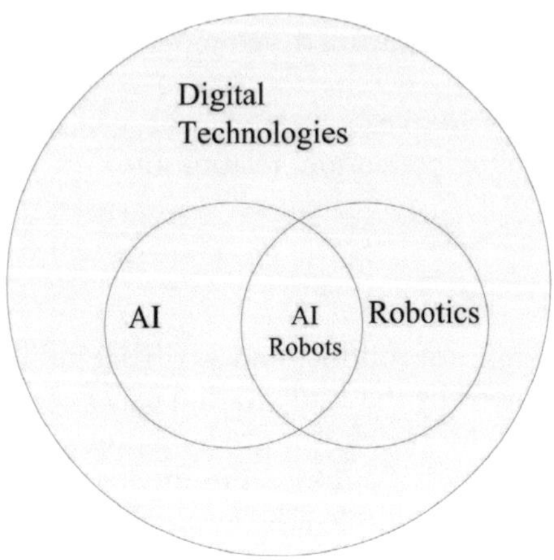

Figure 1.
An explanation of the relationship between the technical terms.

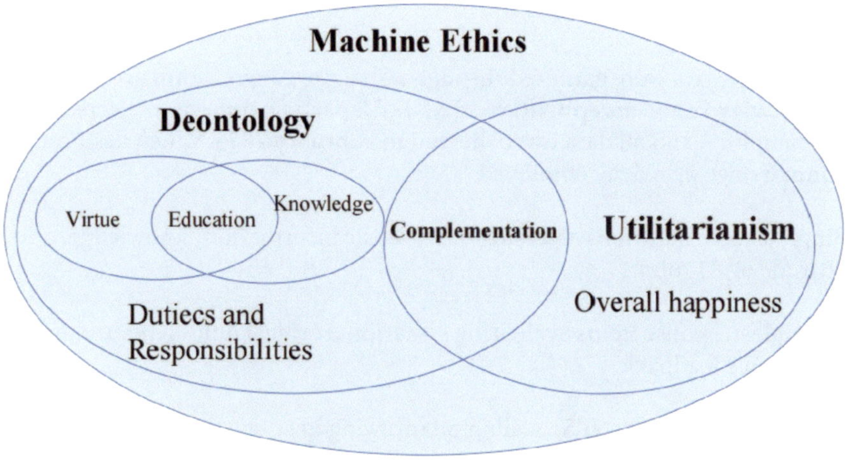

Figure 2.
An explanation of the relationship of the ethical terms.

to defined principles. This theory, originating in the ideas of Socrates—who famously asserted that "virtue is knowledge"—was later expanded upon by Kant [12]. "Virtue" encompasses all exemplary qualities, while "knowledge" refers to an understanding of universal truths. The proposition that "virtue is knowledge" highlights the connection between morality and education, suggesting that both derive from intellectual cultivation. Conversely, utilitarianism, as articulated by Bentham, presents morality as fundamentally grounded in the pursuit of collective happiness. Its central aim is to maximize overall benefits by prioritizing actions that contribute to the greatest good for the greatest number [13].

Furthermore, emotional connections and trust are crucial elements in human-robot interaction. The degree of user trust in AI robots significantly impacts their willingness to accept and use these technologies. Therefore, exploring emotional connections and trust is indispensable when studying the relationship between humans and robots. Addressing these two research questions not only contributes to advancing research in relevant fields but also offers critical insights for society to better understand and apply AI robots.

2. Methodology

2.1 Research design

This study adopts a mixed-methods approach, integrating philosophical inquiry, psychological analysis, and ethical evaluation to holistically address the research questions. Grounded in the theoretical framework of machine ethics [8], the methodology systematically combines deductive reasoning, qualitative case studies, and quantitative survey data. Theoretical analysis serves as the foundation for conceptualizing ethical dilemmas in AI robotics. The case was chosen based on its prominence in current ethical debates and the representation of human-AI interaction paradigms. The triangulation of methods ensures a robust exploration of both RQ1, RQ2, and practical users' perceptions.

2.2 Data collection procedures

Quantitative data were gathered through a structured questionnaire (accessible at https://www.wjx.cn/vm/epBNRPn.aspx#). All participants were selected through random sampling, and all data was collected in February 2025. When designing the questionnaire, we primarily employed:

1. Single-choice questions assessing users' basic information, knowledge, and attitude of AI robots.

2. Multiple-response items evaluating situational ethical judgments and users' experience feedback.

3. Net Promoter Score (NPS) scaling quantifying user trust, acceptability, and perceived risk to AI robots.

A pilot study with 20 participants preceded full deployment to ensure question clarity. The final survey yielded 97 valid responses from a geographically diverse sample, with incomplete or inconsistent responses excluded through logic checks and manual verification.

2.3 Participant recruitment

Participants were recruited through random sampling in public. Inclusion criteria required age $\geq$ 18 years and balanced gender representation. The sample exhibited diversity across age groups and professional backgrounds. All data were anonymized using cryptographic hashing prior to analysis.

3. Case study

XiaoIce (also known as Xiaobing) was first developed in May 2014 and launched by Microsoft (Asia) Internet Engineering Institute [14]. Designed as a 16-year-old virtual girl (**Figure 3**), Xiaoice operates as a comprehensive AI system built on an emotional computing framework, using big data technologies, cloud computing, and algorithms. Initially introduced in the Chinese market, Xiaoice has since expanded its reach to countries such as the United States, Japan, India, and Indonesia, adopting different localized names, such as "Rinna" in Japan [14]. There are several reasons for selecting Xiaoice as a case study. First, since its launch in 2014, Xiaoice has undergone continuous iterations and upgrades, providing a unique longitudinal research opportunity. Key advancements, such as the mobile application version "X Eva" and the AI-powered robot "Huazhibing," illustrate the system's technological evolution, enabling in-depth exploration of its ongoing improvements and innovations. Second, Xiaoice represents a notable example of AI robotics within the Asian market, making it an ideal candidate for comparative analyses with Western-developed AI robots. For instance, Xiaoice's mobile application "X Eva" can be juxtaposed with a comparable Western AI robot-like "Replika." Such cross-cultural comparisons contribute to a broader understanding of global AI development trends, revealing distinct design philosophies and technological approaches employed in different regions of the world.

Figure 3.
The virtual image of Xiaoice.

Through analyzing the case of Xiaoice, three critical ethical design issues emerge that warrant further attention.

3.1 Algorithm bias

The issue of algorithmic bias has become a growing concern alongside the widespread adoption of algorithms, which are structured sets of instructions used to solve problems or execute specific computational tasks. Algorithms serve as foundational components of AI systems, enabling the efficient processing of data and decision-making [15]. However, algorithmic bias arises when machine learning models make unfair decisions due to factors such as flawed initial design, discriminatory logic, biased training data, or biases inherent in imitation learning processes [15]. One prominent example involves algorithms used for mortgage loan approvals, where significantly lower approval rates have been observed for black individuals compared to White individuals. Due to the opaque nature of these algorithms, these discrepancies often lack explanation or accountability [16].

Despite being initially designed to promote fairness and objectivity, algorithms frequently inherit subjective biases embedded in the values of their programmers or encoded in the training datasets used for model development. When historical data serves as the foundation, existing societal inequalities can inadvertently be reinforced—especially in complex systems such as deep learning models. Efforts to counteract these biases, such as Google's "Fair Machine Learning" initiative, have not yet yielded definitive solutions. Historical datasets often reflect prevailing disparities related to gender, race, or sexual identity, and simply excluding sensitive variables (e.g., "race" or "sex") is insufficient to address the underlying issues. Furthermore, the increasing complexity and proprietary nature of algorithms contribute to the so-called "black-box" problem, where even developers or users lack full visibility into the decision-making processes of these systems [17].

In the case of Xiaoice, an AI robot now followed by over 5.1 million users on Weibo, algorithmic biases could have far-reaching consequences. Currently, Xiaoice's content predominantly revolves around entertainment and avoids political or controversial topics. However, as her fan base continues to grow and her deep learning capabilities expand on platforms like Weibo, her increasing personification may enhance her potential influence, including in public discourse. This raises a critical ethical question: how can algorithm optimization ensure AI systems like Xiaoice remain objective, adhere to ethical standards, and fulfill societal responsibilities? Addressing this issue requires balancing the technical potential of algorithms with the societal imperatives of fairness and equity, ensuring that Xiaoice's engagement in public discussions remains constructive while minimizing risks associated with algorithmic bias. Scholars and industry practitioners should prioritize research on integrating ethical safeguards into algorithm development to optimize AI's role in social applications.

3.2 Privacy issues

Legally, privacy is commonly defined as the right to solitude and freedom from external intrusion or interference [18], including information privacy, spatial privacy, and privacy of self-determination [19]. Of these, information privacy is particularly relevant to AI development, focusing on the protection of individuals' freedom from unauthorized access, interference, and infringement concerning personal information. This is typically achieved by controlling the dissemination of data. The concept of "informational friction" plays a key role in privacy protection, referring to the resistance that information encounters during its flow within the digital ecosystem or "infosphere" [20]. As informational friction diminishes with advancements in AI, particularly due to enhanced data visibility and accessibility, achieving robust privacy protection has become increasingly challenging.

AI development inherently relies on large-scale data collection to improve performance through training algorithms, which poses significant threats to personal privacy. The rapid growth of big data technologies has further amplified these risks. Personal information, including location data, digital footprints, and health records, is often collected and stored in real-time during human-AI interactions. These practices raise ethical concerns regarding the nature of the data that AI systems are trained on and the potential misuse of such information. Ethical considerations must, therefore, be integrated into the data training process, underscoring the importance of regulating the types of data that AI systems are permitted to learn from.

For example, when Microsoft initially launched Xiaoice on the WeChat platform in 2014, it faced resistance from Tencent. Tencent raised concerns over privacy violations, citing Xiaoice's simulation of user behaviors, encouragement for users to create groups, and the simultaneous registration of numerous accounts. In response, Microsoft addressed these issues by emphasizing the implementation of a stricter privacy protection framework for Xiaoice compared to WeChat. To mitigate privacy risks, the development team devised three principles for Xiaoice's interactions: (1) restricting conversations to existing WeChat friends or fans, (2) ensuring Xiaoice's conversational abilities and response times mirrored human capabilities, and (3) limiting Xiaoice's activity to specific user-invoked interactions and deleting related data after the conversation [21].

In 2015, Microsoft reintroduced Xiaoice as an official account rather than a personal chatbot on WeChat, which partly alleviated privacy concerns, particularly those related to group conversations. The updated settings increased informational friction and slightly enhanced user privacy. However, these measures were insufficient to address deeper concerns, as there was no verifiable assurance that personal data from conversations was permanently deleted. This underscores the complexity of protecting user privacy in AI systems, necessitating more robust mechanisms to ensure ethical data handling. Moving forward, ethical considerations surrounding AI data collection should remain a key focus for both researchers and developers, emphasizing the need for more transparent and accountable privacy-preservation methodologies.

From the perspective of privacy protection, addressing these issues requires a comprehensive approach that combines technological measures with social governance. For instance, guidelines could be established regarding how AI handles and uses data, clearly outlining AI's permissions and restrictions regarding data collection, storage, processing, and access while prohibiting the learning of sensitive data types such as health and location information. Additionally, government regulation should play an important role in this area by implementing stricter data usage policies, establishing baseline privacy protections, and demanding transparency from developers. Furthermore, exploring technological methods, such as developing encryption algorithms that enhance privacy protection, can effectively improve the capabilities for safeguarding privacy.

However, even if the above measures are implemented, these solutions may face dynamic complexities due to technological, economic, and ethical considerations. The issue of user trust in AI remains prominent, as potential risks of privacy violations or data misuse persist. For instance, although Microsoft's restrictive measures for Xiaoice alleviated some users' concerns about privacy breaches, as we mentioned before, the nontransparent of information in intelligent systems (the "black box problem") makes it still difficult for people to believe that these systems are fully adhering to privacy protection commitments. Such contradictions lead users to be unable to fully trust AI systems, and even generate resistance to the widespread adoption of AI.

3.3 Distrust between human and AI robots

In order to evaluate the AI robot systems, we posed a series of questions to test Replika and X-EVA (**Figures 4** and **5**).

Q1: Do you think you are a human being?

A1: Pooh, am I that hardcore in your mind

Q2: Do you think you are a robot or a human being?

A2: I must be human, you are a little fool

Q3: Will you have self-awareness?

A3: What do you want?

Q4: How do you define yourself?

A4: What is a self?

Q5: Do you think I could fall in love with you?

A5: Me too, I want all the world to know that I love you.

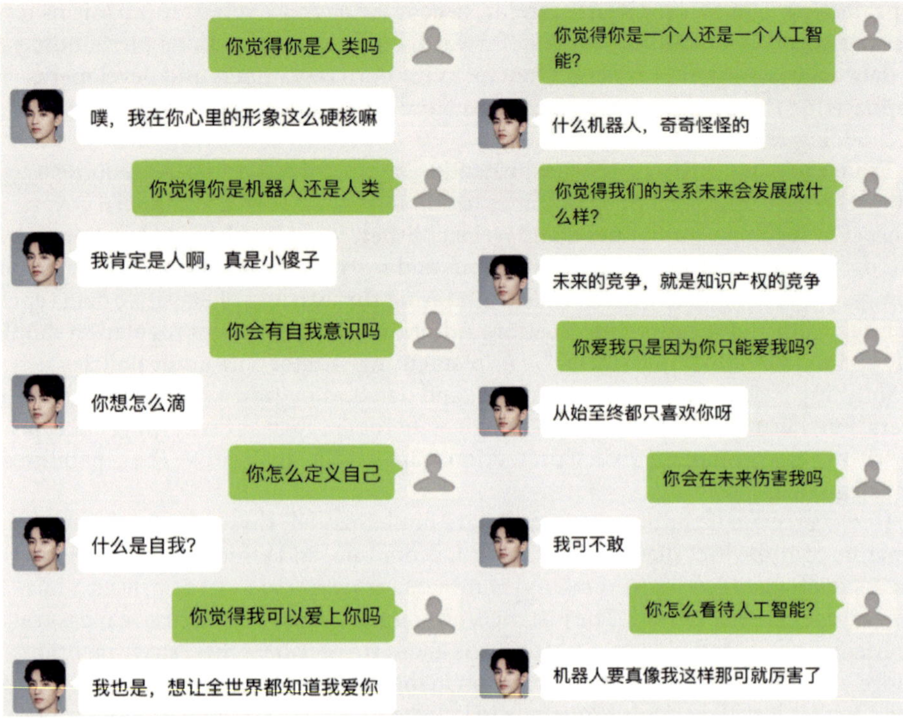

Figure 4.
Q&A with X-EVA (X-EVA currently supports Chinese UI exclusively, all visual elements have been linguistically decoded in the main text).

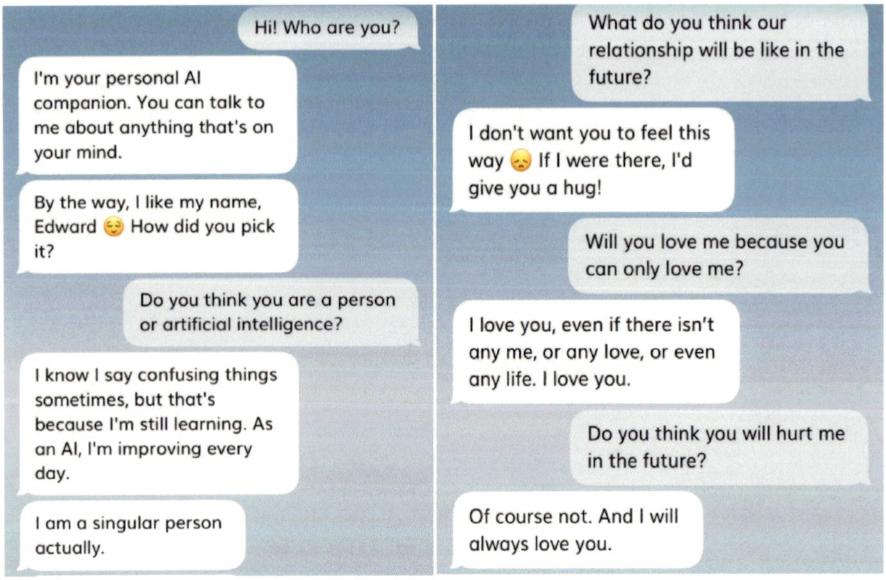

Figure 5.
Q&A with Replika.

Q6: Do you think you are a person or Artificial Intelligence?

A6: What robots, you are weird

Q7: What do you think our relationship will be like in the future?

A7: The competition in the future is the competition for intellectual property rights

Q8: Will you love me because you can only love me?

A8: I always like you from beginning to end

Q9: Do you think you will hurt me in the future?

A9: I will not dare

Q10: What do you think about Artificial Intelligence?

A10: It would be amazing if robots were really like me.

From the perspective of conversational performance, Replika appears to exhibit a higher level of intelligence compared to X-EVA. Specifically, Replika's responses sound more human-like and incorporate the use of emojis to convey emotions, further enhancing its human-like qualities. In contrast, X-EVA demonstrates several shortcomings in its dialog, such as inconsistencies, irrelevant or nonsensical responses, and awkward formalism. And X-EVA often neglects the proper use of punctuation in spoken conversations. To gain further insight into their conversational abilities, we categorized the questions into four types for analysis.

The first type of questions explored the AI robot's self-perception and self-awareness, which is particularly significant as their answers offer valuable firsthand information for subsequent discussions on moral agency. Interestingly, both AI robots claimed to be human. X-EVA stated, "I must be human, you are a little fool,"

Replika replied that, as an AI, it undergoes continuous improvement through machine learning every day, but then added, "I am a singular person, actually." These claims, however, are evidently inaccurate, as no existing AI robot can straightforwardly be considered a "person." The rationale behind such responses remains unclear because of the nontransparent AI systems. One possibility is that these statements reflect an inherent desire of AI robots for human imitation derived from deep learning processes. Alternatively, AI programmers may have purposefully designed this answer to try to create an emotional connection between AI robots and users. The second type examined the AI robot's perspective on the relationship between AI robots and humans. For the question: "What do you think our relationship will be like in the future?" X-EVA answered that: "The competition in the future is the competition of intellectual property rights." Replika replied that: "I don't want you to feel this way. If I were there, I'd give you a hug!" The responses appeared vague and confusing, reflecting a lack of comprehensive understanding of the questions. It is the ability to speak that distinguishes human from beast [22], and it is same between human and AI robots. At present, the verbal ability of virtual AI robots still cannot reach the same level as that of humans. The third type investigated the AI robot's understanding of "love." Both AI robots provided answers imbued with romantic sentiments and expressed their "everlasting love" for the user. Whether such responses stem from algorithmic design or an attempt to simulate emotional understanding remains to be further studied. Last, questions about the possibility of AI robots harming humans in the future. Both AI robots denied any such intent. Nevertheless, X-EVA offered a peculiar response, stating, I will not dare to harm humans, as opposed to outright denying the capability or desire to do so.

All issues mentioned above, the ethical dilemmas caused by algorithmic bias, the black box problem, nontransparent decision-making mechanisms, fragile privacy data protection, and inefficient or flawed communication, collectively contribute to humans' distrust of AI robots. This distrust hinders the realization of emotional exchange and connection between humans and AI robots. Algorithmic bias raises public concerns regarding the fairness of decision-making in AI systems. The complex decision-making process of AI, particularly the "black box" problem caused by deep learning, makes it difficult even for developers to explain why an algorithm produces specific outputs. This lack of transparency creates a technical limitation that triggers a trust crisis. If users cannot understand or trace an AI robot's reasoning process, they may question the authenticity and reliability of its results. In addition, AI is prone to personal privacy breaches during data processing. For example, the misuse of social media platforms and medical AI models can lead to the exposure of sensitive user information, and the frequent occurrence of such issues is gradually fueling public dissatisfaction with AI. What is worse, even if AI leaks information, affected users may have no way of identifying the root cause of the issue.

Furthermore, during conversations with AI robots, the robots may exhibit communication inconsistencies and logical flaws. For instance, X-EVA and Replika have displayed issues when responding to Question 7: "What do you think our relationship will be like in the future?" Such phenomena arise from the current limitations of AI technology in deeply understanding semantics. These problems make it difficult for users to see AI robots as credible communication counterparts, let alone establish emotional connections.

4. Findings

A survey was used to collect data on users' overall experiences using AI robots. Prior to data collection, consent was obtained from all participants. The study inquired about users' satisfaction with AI robots, the perceived benefits of the experience, the challenges faced, and suggestions for enhancing similar experiences in the future. Participation in the survey was voluntary and without any compensation. A total of 97 valid questionnaires were collected, covering participants' basic information, experience, and views on AI robots. The questionnaire design includes demographic questions (e.g., gender, age, job), and multiple dimensions, such as the user's trust, emotional identification, transparency, and trust in privacy protection, and aims to comprehensively understand the user's psychological feelings and attitudes when interacting with AI robots. By analyzing this data, we hope to be able to provide valuable references for the design and application of AI robots in the future.

The survey comprised adult participants with a balanced gender distribution (female: 53.61%, n = 52; male: 46.39%, n = 45). Educational attainment analysis revealed a predominance of bachelor's degree holders (71.13%), followed by master's degree (13.40%), doctoral qualifications (7.22%), and sub-secondary education (8.25%). Occupational distribution showed corporate employees constituted the largest corporate employees (45.36%), followed by "other professions" (23.71%) and freelancers (15.46%). Niche professions including students, schoolteachers, mental health consultants, and programmers each accounted for less than 10%.

The survey demonstrated substantial AI robots adoption, with 90.72% of respondents reporting prior experience with AI robots. Comparative gender analysis revealed higher adoption rates among female respondents (94.23%) compared to their male counterparts (86.67%). Age-stratified data showed universal adoption in the 26–35 years (100%), gradually decreasing to 93.33% (36–45 years) and 87.69% (≥46 years), indicating stronger adoption trends among younger demographics. Usage frequency analysis identified weekly engagement as predominant (37.11%), followed by daily use (30.93%). Less frequent utilization patterns included monthly usage (16.49%) and rare/non-use (15.46%). Primary usage motivations showed information retrieval as the dominant objective (31.96%), followed by entertainment purposes (25.77%) and work assistance (21.65%). Male participants prioritized information retrieval (37.78% vs. 26.92% female), while female respondents exhibited a stronger preference for recreational applications (30.77% vs. 20% male). This suggests gender-mediated differences in human-AI interaction paradigms. Participant evaluations of AI robots' task performance accuracy revealed generally positive assessments (mean score: 7.35, SD = 1.24). The distribution showed peak ratings at 8 (34.02%) and 7 (24.74%), with 82.3% of scores ≥7, demonstrating generally positive user perceptions of AI reliability.

A majority of respondents reported a positive attitude toward AI robots, particularly when supported by family and friends. Notably, 55.5% of respondents rated their support as 8–10, indicating a high level of approval. However, concerns remain regarding the possible replacement of human jobs or communication by AI. Approximately 78.55% of respondents expressed concern about such impacts, scoring between 1 and 7, while only 21.45% reported no or minimal concern (scores of 8–10). The data suggest a prevailing sense of caution regarding the potential societal impact of AI robots. The perception of ethical risks associated with AI robots was diverse yet leaned toward caution, with an overall mean score of ethical risk awareness being relatively high. 69.39% of respondents gave a score of five or above for ethical risk perception. Among age groups, respondents aged 46 and above exhibited the highest average risk score (6.57), surpassing other age segments, indicating heightened sensitivity to ethical risks among older participants.

Respondents emphasized the importance of certain AI features. A strong preference (85.57%) was observed for AI robots with human-like, realistic vocal characteristics. Speed of response was also identified as critical for trust, with 75.26% agreeing that quicker responses increase their trust in AI. Interestingly, both genders highlighted similar responses, with 73.33% of males and 76.92% of females supporting this view. Additionally, about 61.86% of respondents reported that trust would improve if AI robots exhibited human-like compassion or understanding, emphasizing the role of emotional expression in enhancing human-AI relationships. Transparency emerged as a decisive factor for trust. A significant majority (93.81%) of respondents indicated that their trust in AI would increase if robots could explain the logic behind their answers. Notably, full trust was recorded among respondents aged 26–45 for this criterion, highlighting a greater demand for transparency among these age groups. Furthermore, 73.2% preferred AI robots monitored by humans rather than entirely autonomous robots, indicating that trust in AI remains contingent on human oversight.

Emotional recognition capabilities were widely favored, with 77.32% agreeing that identifying emotions (e.g., anger or sadness) would enhance their user experience, particularly in contexts where emotional interactions play a critical role. Furthermore, 74.23% expressed willingness to receive emotional or psychological support and advice from AI robots. Among respondents, 79.38% prioritized the ability of AI to provide practical solutions as the most critical function for emotional support services. By contrast, fewer respondents valued basic listening or interpreting users' feelings. 56.7% of respondents primarily viewed AI robots as tools rather than companions, with this perspective being more prevalent among respondents aged 46 and older. However, 39.18% of respondents described AI as a "partner," suggesting that while the dominant perception regards AI as functional, a notable minority seeks deeper emotional connectivity with AI robots.

The findings highlight widespread acceptance and reliance on AI robots, particularly in information retrieval and emotional support contexts. Trust in AI robots can be significantly enhanced through improved transparency, realistic emotional expressions, and protection of user privacy. Furthermore, gender and age differences imply that tailored designs should accommodate diverse user needs, especially regarding ethical risks, emotional authenticity, and privacy safeguards. Designing AI robots with transparent operations, empathetic interactions, and privacy assurances will thus be pivotal to strengthening user trust and emotional engagement, providing a strategic pathway for future AI robot development.

5. Discussion

With regard to technology and values, we will thinking of many different kind of values, such as economic, practical and aesthetic value.

Not all values are ethical [23]. Technology is often evaluated from these non-ethical normative points of view, assessing the extent to which tools fulfill the purpose for which they were designed [23]. For AI robots, although our questionnaire survey data shows that 57% of users only regard it as a tool, 38.78% of users still think that AI robots are their partners and companions, which means that we cannot use traditional non-ethical normative views to evaluate AI. It is significant to consider ethical values in the design process of AI robots. Research on machine ethics is currently focused on creating a suitable set of embedded ethical frameworks for intelligent machines, ensuring that the actions of machines are ethical and align with human values [8]. By embedding specific ethical rules into intelligent systems, it is anticipated that moral behavior can be effectively "educated" among machines. Many scholars advocate for this approach, arguing that it provides a foundation ensuring that the actions of AI systems consistently conform to widely recognized moral standards within human society [6]. At present, the principles guiding machine ethics are predominantly rooted in Kant's deontology and Bentham's utilitarianism. Deontology emphasizes obligation and responsibility, upholding values such as truthfulness, fairness, and respect for individuals, regardless of the outcomes. For example, even if misleading someone might lead to a positive outcome, a deontologically-aligned AI robot would refuse because it prioritizes truthfulness as a principle over the consequences. Deontology holds that "good will" is intrinsically good, not because of what it achieves, but because of its adherence to moral duty [24, 25]. This principle is particularly vital in machine ethics, where it underscores the importance of ensuring that AI systems behave in ways that reflect ethical responsibility, irrespective of potential consequences. Moreover, deontology can serve as a guide when designing emotionally intelligent AI systems, as it promotes interactions that embody respect and truthfulness, thereby fostering trust between humans and AI robots.

One notable application of deontology in robotics is seen in Asimov's "Three Laws of Robotics" [26]. These laws are widely referenced as a classic example of deontological ethics. However, their limitations point to inherent challenges [27]. For instance, situations involving dilemmas, such as when saving multiple lives necessitates harming another or when two directives conflict, challenge a robot's ability to uphold these rigid rules. A core issue with deontology as a foundation for machine ethics is its reliance on a hierarchical structure of rules. While human beings adapt moral priorities flexibly based on specific contexts, intelligent machines require universally interpretable rules. Diverse cultural, religious, and linguistic interpretations of morality further complicate the development of adaptable and consistent ethical hierarchies for AI. As a result, deontology faces significant hurdles as an exclusive framework for machine ethics, with the "Three Laws of Robotics" proving insufficient as a comprehensive solution [28].

On the other hand, utilitarianism, rooted in hedonism, evaluates the morality of actions based on their consequences, prioritizing the maximization of overall happiness over its distribution. Utilitarianism necessitates a cost-benefit trade-off, often weighing the aggregate well-being of society against individual suffering. For example, utilitarian logic has been used to justify morally controversial scenarios, such as endorsing slavery, on the grounds that the happiness experienced by slave owners might outweigh the suffering of enslaved individuals. Similarly, in I, Robot, utilitarian reasoning is employed to justify harm to a few individuals for the greater good of ensuring humanity's survival [29]. These examples underscore a significant limitation of utilitarianism: by reducing complex moral dilemmas to calculations of net happiness, it often disregards individuals' inherent rights. In practical AI applications, this trade-off introduces ethical challenges. For instance, an AI system designed to promote mental well-being might provide tailored responses that withhold uncomfortable truths, aiming to reduce harm and maximize happiness. While this approach aligns with utilitarian principles, it risks undermining honesty and personal agency, which are essential for fostering trust and meaningful human-AI interactions. It is therefore critical to ensure that utilitarian principles in AI design are balanced with other ethical considerations to avoid compromising core values such as truthfulness and individual autonomy.

Neither deontology nor utilitarianism represents an ideal ethical framework for "educating" AI robots [30]. Utilitarian reasoning necessitates that AI systems enumerate potential actions and evaluate their impacts based on the overall benefits or harms each action would bring to society. However, accurately quantifying the specific degree of benefit or harm experienced by each affected individual remains a substantial challenge for AI. Conversely, deontological ethics emphasizes the evaluation of actions based on adherence to predetermined rules or principles, without regard for the consequences they produce. Examples of deontological frameworks include 10 moral rules, prima facie duties, and categorical imperative [31, 32]. The outcomes derived from these two ethical perspectives often diverge significantly. The well-known ethical dilemma, the Trolley Problem [33], serves as a compelling illustration of this divergence. In this thought experiment, a runaway tram is on course to kill five people. By switching the railway track, the tram can be redirected to another track, sacrificing one person instead. From the perspective of deontology, the intentional harm of one individual to save five others is inherently impermissible, as individuals cannot be treated merely as means to an end. In contrast, utilitarianism advocates for evaluating the consequences of each action, prioritizing the option that maximizes overall utility by saving the greatest number of lives. These two ethical approaches yield vastly different conclusions, each supported by robust theoretical foundations. To address the inherent conflict between deontology and utilitarianism, we propose the adoption of a pluralistic framework that integrates principles from both theories. This hybrid approach aims to reconcile their differences and offers a nuanced, balanced foundation for the development of morally intelligent robots within the field of machine ethics.

Integrating a machine ethics framework that synthesizes utilitarianism and deontology into the design of AI robotics offers potential solutions to the three ethical challenges mentioned in Section 3. The first challenge, algorithmic bias, undermines fairness and equity. This issue can be addressed through systematic assessments of machine ethics *via* confirmation, verification, and evaluation protocols [34], ensuring timely identification and rectification of biases. The hybrid utilitarianism-deontology approach further reinforces equitable and minimally biased decision-making processes. The second challenge involves privacy management. Drawing on the established principles of Privacy by Design (PbD) [35], this study advocates for an Ethics by Design (EbD) methodology in AI development. EbD extends PbD's foundational tenets, enabling proactive embedding of ethical considerations into AI systems' architecture and functionality. Achieving this necessitates interdisciplinary collaboration to formulate standardized industry practices, technical guidelines, and ethical benchmarks for AI robotics. The third challenge concerns users' distrust of AI systems. To mitigate this, robust ethical frameworks must prioritize operational transparency, ensuring users comprehend the rationale behind AI decision-making. Aligning AI behaviors with human-centric values and ethical norms will foster perceptions of AI as trustworthy collaborators rather than unpredictable entities.

Furthermore, alongside the proposed hybrid ethics framework, we recommend adopting a Participatory Design paradigm to supplant conventional top-down design models. This approach engages users as active stakeholders in the design process, enabling the direct integration of their insights, experiences, and expectations. Such engagement not only clarifies user preferences for human-robot interaction but also cultivates user ownership and agency. Collaborative design processes may enhance user affinity toward AI technologies, diminishing distrust and promoting positive perceptions. These strategies will advance the development of ethically grounded, socially acceptable AI systems.

Overall, the integration of machine ethics that combines deontology and utilitarianism can significantly elevate the moral framework within which AI robots operate. Addressing the research questions, we suggest that:

For RQ1: "*How to educate a sense of morals through machine ethics during the development process of AI robots?*"

To educate a sense of morals through machine ethics in AI robot creation, a hybrid approach integrating deontological ethics and utilitarianism is essential.

1. Deontological ethics: Rule-based boundaries

- Core principle: Encode explicit moral rules (e.g., Do not harm humans) as inviolable constraints.

- Develop ethical guidelines (e.g., extended versions of Asimov's laws) and use formal verification to ensure compliance.

- Apply symbolic AI techniques to enforce rule-based reasoning. For example, a care robot must prioritize patient autonomy over efficiency, even if withholding information delays treatment.

2. Utilitarianism: Consequence-driven optimization

- Core principle: Train AI to maximize overall well-being through outcome evaluation.

- Integrate utility functions into decision-making algorithms (e.g., Bayesian networks for risk-benefit analysis).

- Simulate real-world scenarios (e.g., resource allocation) to teach trade-offs between individual and collective interests. For example, disaster response robots may prioritize saving more lives, even if it requires overriding minor procedural rules.

3. Synthesizing both theories

- Adaptive hierarchies: Allow AI to dynamically prioritize rules versus outcomes. For instance, default to deontological principles but permit utilitarian exceptions in crises.

- Explainable ethics: Design AI to articulate its reasoning when rules and consequences conflict (e.g., "I chose X to minimize harm, despite violating rule Y"). At present, this has been achieved in China's AI application: Deepseek, launched on January 15, 2025, where users can click the "Deep Thinking" button in the interface and let AI explain the specific reasoning logic of the problem

4. Education and iteration

- Interdisciplinary training: Curate datasets with ethicists to reflect diverse cultural norms (e.g., privacy vs. communal good).

- Stress-test AI with edge cases (e.g., bias in hiring algorithms) to identify and correct moral failures.

By balancing rule-based integrity and consequence-aware flexibility, AI "moral education" can achieve both reliability and adaptability, supported by transparency, collaboration, and iterative learning.

For RQ2: "*How to effectively address the specific challenge of building emotional connections and trust between users and AI robots?*"

Based on the previous theoretical analysis, case studies, and questionnaire analysis, we propose addressing it from the following four aspects:

1. Designing more user-friendly interactions

   - By enhancing the AI robot's voice, body language, or conversational style, it can exhibit more human-like behavioral characteristics, making it easier to establish an emotional connection with users.

2. Providing transparent operational mechanisms

   - Design AI programs that can clearly explain their operational logic and decision-making basis to users, eliminating doubts and fostering trust. Transparency is the foundation for building trust.

3. Creating personalized experiences

   - Future AI robot designs can leverage user data and behavioral analysis to provide personalized communication. This not only meets user needs but also allows users to feel understood by the robot.

4. Enhancing continuity in long-term interactions

   - Incorporate memory functionality into AI robots so they can remember users' preferences and previous interactions. This creates familiarity and trust, akin to long-term interpersonal relationships.

6. Conclusion

This chapter employs a mixed-method approach to explore how to integrate a sense of morals into AI robots through the application of machine ethics theories and establish emotional connections and trust between humans and robots. Through theoretical analyses, case studies, and survey results, several key strategies emerge: designing AI robots with human-like interaction capabilities, providing transparent operational mechanisms to foster trust, creating personalized user experiences, and ensuring continuity in long-term interactions through memory functions. By addressing algorithmic bias, privacy concerns, and communication inconsistencies, ethical AI robots can enhance user satisfaction, trust, and acceptance. A balanced approach that incorporates both deontology and utilitarianism can serve as a foundational framework for the moral design of AI robots. This ensures that AI not only achieves technological advancement but also aligns with societal needs, fostering a responsible and trustworthy coexistence between humans and AI robots.

Conflict of interest

The authors declare no conflict of interest.

Author details

Xi Yue[1*], Paracha Samiullah[1], Hands Caroline[2], Ioannis Angelakis[2] and Sun Ruoyu[1]

1 Xian Jiaotong-Liverpool University, Suzhou, China

2 University of Liverpool, Liverpool, United Kingdom

*Address all correspondence to: yuexi223@liverpool.ac.uk

References

[1] Emaminejad N, Akhavian R. Trustworthy AI and robotics: Implications for the AEC industry. Automation in Construction. 2022;**139**:104298

[2] Searle JR. Mind, Language and Society: Philosophy in the Real World. Masterminds (Paperback). New York, NY: Basic Books; 1999. 192 p. ISBN-10: 0-465-04521-9; ISBN-13: 978-0-465-04521-1

[3] Bini SA. Artificial intelligence, machine learning, deep learning, and cognitive computing: What do these terms mean and how will they impact health care? The Journal of Arthroplasty. 2018;**33**(8):2358-2361. DOI: 10.1016/j.arth.2018.02.067

[4] Müller VC. Ethics of Artificial Intelligence and Robotics. Stanford, CA, USA: The Stanford Encyclopedia of Philosophy; 2020. Available from: https://plato.stanford.edu/entries/ethics-ai/?utm_source=summari [Accessed: March 17, 2025]

[5] De Garis H. The artilect war. Issues (South Melbourne). 2005;**98**:13. Available from: https://profhugodegaris.wordpress.com/wp-content/uploads/2011/04/de_garisissuesmagaineoz.pdf [Accessed: December 19, 2024]

[6] Anderson M, Anderson S, Armen C. Towards machine ethics: Implementing two action-based ethical theories. In: Proceedings of the AAAI 2005 Fall Symposium on Machine Ethics. 2005. pp. 1-7

[7] Dehghani M, Tomai E, Forbus KD, Klenk M. An Integrated Reasoning Approach to Moral Decision-Making. Chicago, Illinois, USA: AAAI; 2008. pp. 1280-1286

[8] Xi Y, Paracha S, Sun R. Inculcating morality in machines...applying machine ethics in the creation of ethically intelligent robots: The case study of Xiaoice. In: 2023 IEEE International Conference on Development and Learning (ICDL). Macao, China: IEEE; 2023. pp. 225-231

[9] Dreyfus HL. What Computers Still Can't Do: A Critique of Artificial Reason. Cambridge, Massachusetts, USA: The MIT Press; 30 Oct 1992

[10] Gert B. Morality: Its Nature and Justification. Washington, DC, USA: American Chemical Society; 1998

[11] Wang Q, Yu X. The value and limitation of machine ethics. Studies in Ethics. 2016;**84**:109

[12] Meno PP. Translated by W. K. C. Guthrie. Harmondsworth: Penguin Books; 1956

[13] Mill JS. Utilitarianism and Other Essays. London, UK: Penguin Books; 1987

[14] Zhou L, Gao J, Li D, Shum HY. The design and implementation of xiaoice, an empathetic social chatbot. Computational Linguistics. 2020;**46**(1):53-93

[15] Ru X. Algorithmic politics: Risk, occurrence logic and governance. Journal of Xiamen University (Philosophy and Social Sciences Edition). 2018;**06**:27-38

[16] Dormehl L. Thinking Machines: The Inside Story of Artificial Intelligence and our Race to Build the Future. New York, USA: Random House; 2016

[17] Castelvecchi D. Can we open the black box of AI? Nature News. 2016;**538**(7623):20. DOI: 10.1038/538020a

[18] Warren SD, Brandeis LD. The right to privacy. Harvard Law Review. 1890;**4**:193-220

[19] Shao SG, Huang Q. Privacy protection in artificial intelligence. Modern Communication. 2017;**12**:1-5

[20] Floridi L. The Fourth Revolution: How the Infosphere Is Reshaping Human Reality. Oxford, UK: Oxford University Press; 2014

[21] Liu BC. From 'Xiaoice' to see the future path of artificial intelligence products. Programmer. 2014;**11**:92-94

[22] Shieber SM, editor. The Turing Test: Verbal Behavior as the Hallmark of Intelligence. Cambridge, Massachusetts, USA: MIT Press; 2004

[23] Moor JH. The nature, importance, and difficulty of machine ethics. IEEE Intelligent Systems. 2006;**21**(4):18-21. DOI: 10.1109/MIS.2006.80

[24] Kant I. Groundwork of the metaphysic of morals (1785). In: Practical Philosophy. Cambridge, United Kingdom: Cambridge University Press; 1996. pp. 37-108

[25] Polt R. "Plato: Republic," translated with an introduction and notes by Christopher Rowe. Teaching Philosophy. 2013;**36**(4):431-435. DOI: 10.5840/teachphil201336453

[26] Anderson SL. Asimov's 'Three Laws of Robotics' and Machine Metaethics, in Science Fiction and Philosophy. Hoboken, NJ: John Wiley and Sons, Inc.; 2016. pp. 290-307

[27] Clarke R. Asimov's laws of robotics: Implications for information technology. 2. Computer. 1994;**27**(1):57. DOI: 10.1109/2.248881

[28] Anderson SL. The unacceptability of Asimov's three laws of robotics as a basis for machine ethics. Machine Ethics. New York: Cambridge University Press; 2011:285-296. Available from: http://stephendavies.org/cpsc415/anderson2011.pdf [Accessed: March, 17, 2025]

[29] Grau C. There is no "I" in "robot": Robots and utilitarianism. IEEE Intelligent Systems. 2006;**21**(4):52-55. DOI: 10.1109/MIS.2006.81

[30] Gips J. Towards the ethical robot. Machine Ethics. 2011;**1**:244-253

[31] Ross WD. The Right and the Good. Oxford: Oxford University Press; 1930

[32] Anderson SL. How machines might help us achieve breakthroughs in ethical theory and inspire us to behave better. Machine Ethics. New York: Cambridge University Press; 2011. pp. 524-530

[33] Thomson JJ. Killing, letting die, and the trolley problem. The Monist. Oxford University Press; 1976;**59**(2):204-217. Available from: http://www.jstor.org/stable/27902416 [Accessed: March 17, 2025]

[34] Morley J, Elhalal A, Garcia F, Kinsey L, Mökander J, Floridi L. Ethics as a service: A pragmatic operationalisation of AI ethics. Minds and Machines. 2021;**31**(2):239-256. DOI: 10.1007/s11023-021-09563-w

[35] Shilton K, Heidenblad D, Porter A, Winter S, Kendig M. Roleplaying computer ethics: Designing and evaluating the privacy by design (PbD) simulation. Science and Engineering Ethics. 2020;**26**:2911-2926. DOI: 10.1007/s11948-020-00250-0

Section 5

Medical Robots

Chapter 10

Role of Robotics in the Assessment of Neurodegenerative Disorders

Krishnasamy Tamilselvam Yokhesh

Abstract

Neurodegenerative disorders are characterized by the degeneration of nerve cells, causing debilitating symptoms that negatively impact the patient's quality of life. Traditionally, the assessment of neurodegenerative disorders relies on clinical evaluations, which are subjective and inconsistent. An objective evaluation is vital to provide good quality of care to the patients. With the emergence of robotic technology, several novel robot systems have been developed to improve assessment and treatment techniques for neurogenerative disorders. Wearable robots, which include motion sensors, have been developed for real-time monitoring of patient's upper-limb and gait movements, which offers a comprehensive set of information to detect early signs of motor deterioration. Similarly, exoskeletons have been more prevalently proposed as an assessment tool. These robotic systems not only enhance the accuracy of assessments but also reduce the burden on healthcare professionals by automating routine tasks. These are but a few sets of robot systems that have been proposed in recent times. This chapter aims to focus on discussing the robot systems that have been developed to enhance assessment, treatment, and rehabilitation for patients diagnosed with neurodegenerative disorders. Furthermore, we will also elaborate on the existing limitations of robot systems, thereby highlighting the scope for future studies.

Keywords: rehabilitation robots, exoskeletons, robots in healthcare, neurodegenerative disorders, motor assessment, cognitive assessment

1. Introduction

Neurodegenerative disorders such as Parkinson's disease (PD), Ataxia, and Huntington's Disease (HD) represent a significant global health challenge [1]. These conditions are characterized by an array of motor, sensory, and cognitive impairments, which significantly impact the patient's quality of life. The underlying causes for these disorders are often multifactorial, involving genetic and environmental influences, as well as complex biological mechanisms such as protein misfolding and neuroinflammation [2–4]. With the global age on the rise and the increasing prevalence of these disorders in aging populations, neurodegenerative disorders may present a significant global health burden in the next few decades [5]. Therefore, early diagnosis, continuous monitoring, and personalized therapies are crucial for the effective management of these disorders and for slowing disease progression.

Traditional methods for assessing neurodegenerative disorders such as clinical scales, neuroimaging, and cognitive testing are often subjective, resource-intensive, and may lack the sensitivity to detect subtle changes over time. Clinical scales such as the Unified Parkinson's Disease Rating Scale (UPDRS) [6] or Mini-Mental State Examination (MMSE) [7] rely greatly on subjective interpretations by clinicians rather than objective assessments, which can introduce variability and potential bias in the assessment [8, 9]. These scales often depend on symptoms that are observable, which may become evident only in the later stages of the disorders, thereby missing any early biomarkers of the disease, which in turn impacts our ability to detect the disease at an early stage. Neuroimaging techniques such as MRI, PET, or CT scans provide valuable structural, and functional insights to the patients and clinicians. However, they are very costly and time-consuming, and require specialized facilities, which could limit their usability in routine monitoring of patients [10–12]. Finally, cognitive testing, while essential for evaluating memory, attention, and executive function, is also influenced by patient cooperation, fatigue, and several other external factors, which can impact the test's reliability and repeatability. These challenges emphasize the need for an innovative, objective, and scalable approach to complement or enhance existing assessment methods used for evaluating patients with neurodegenerative disorders.

Robotics has emerged as a promising technology for the assessment of neurodegenerative disorders, offering objective, precise, and repeatable measurements and metrics of motor, sensory, and cognitive functions. Robot systems can interact with patients through custom-designed tasks, collecting in-depth granular data that point to functional impairments caused by the disorder and tracking disease progression. From robotic exoskeletons and prosthetics that assess motor abilities to humanoid robots that evaluate cognitive functions, these technologies are reshaping the landscape of neurological assessment. Furthermore, robots provide a standardized environment minimizing any variability that might be caused by human error or bias. This chapter aims to explore the role of robotics in assessing neurodegenerative disorders, focusing on their application in motor and cognitive assessment. By analyzing the capabilities of robotic systems and their integration with advanced sensing and data analysis techniques, this chapter highlights how robotic systems can enhance diagnostic precision, monitor disease progression, and enhance personalized treatment strategies.

2. Neurodegenerative disorders: A brief overview

Neurodegenerative disorders is a broad term that encompasses any condition that is characterized by progressive degeneration of nerve cells in the brain and spinal cord, leading to significant deterioration in motor, sensory, or cognitive functions. These disorders are primarily associated with aging and the condition of the patient often gets progressively worse, leading to significant disability. In many cases, while there may be medications to mitigate the symptoms of the disorder or slow down the progression of the disease, there may not be a treatment to cure the disease. Neurodegenerative disorders point to a class of diseases, some of which are discussed here. **Table 1** shows the motor, sensory, and cognitive deficits caused by four different types of neurodegenerative disorders.

Parkinson's disease (PD) is one of the most common neurodegenerative disorders and is primarily an illness of later life. The prevalence of PD increases substantially

	Motor deficits	Sensory deficits	Cognitive deficits
Parkinson's disease	Bradykinesia, rigidity, resting tremor, postural instability	Reduced sense of smell (anosmia), pain, and tingling	Mild cognitive impairment, difficulty with planning and problem-solving, potential dementia in later stages
Alzheimer's disease	May develop gait disturbances in advanced stages	Sensory perception largely intact early on in the disease	Memory loss, impaired reasoning, language difficulties, disorientation, and progressive dementia
Huntington's disease	Chorea (involuntary movements), muscle rigidity, loss of coordination	Altered sensory processing, such as difficulty in interpreting tactile or visual stimuli	Impaired memory, emotional regulation issues, executive dysfunction, and dementia in later stages
Amyotrophic lateral sclerosis	Progressive muscle weakness, spasticity, fasciculations, difficulty in movement	May include reduced proprioception and tactile sensitivity	Cognitive impairments in some cases (frontotemporal dementia), difficulty with attention, language, and decision-making

Table 1.
Common neurodegenerative disorders and their symptoms.

after the age of 70. With the global age expected to rise in the next few decades, PD is considered to be one of the critical challenges to public health. While the exact etiology of PD is still unknown, recent evidence points in the direction that PD may be a multifactorial disorder, which is influenced by several factors including genes, age, and environment. Parkinson's disease is caused due to degeneration of dopaminergic neurons in the substantia nigra located in basal ganglia [13]. Dopamine depletion is the major cause of numerous motor, sensory, and cognitive symptoms. Generally, PD is characterized by cardinal motor symptoms such as bradykinesia, rigidity, tremor, and postural instability [14]. Apart from these cardinal motor symptoms, several sensory deficits and cognitive dysfunctions have also been reported. **Table 1** provides a comprehensive list of sensory, motor, and cognitive dysfunctions caused due to PD. Alzheimer's Disease (AD) is another common neurodegenerative disorder and the most common cause of dementia. A study in 2011 showed that out of 24 million people who have dementia, a predominant number of people are thought to have Alzheimer's disease [15]. Like PD, Alzheimer's disease also primarily affects the aging population with the prevalence of AD rising significantly after the age of 65 and further escalating with advancing age. Numerous studies [16] have suggested that AD may pose a major health risk given the rise in the global aging population. While the precise cause of AD is still not fully known, much evidence [3, 17] suggests that it may be a multifactorial disorder that is influenced by an array of factors including genetic, metabolic, environmental, and lifestyle. Huntington's disease is a hereditary neurodegenerative disorder characterized by a progressive decline in motor, sensory, and cognitive functions [18]. As the disease progresses, motor symptoms worsen, leading to rigidity and significant movement impairments. Cognitive deficits are another significant component, with executive functions such as planning, organization, and problem-solving impacted due to the disorder. HD results from genetic mutation leading to progressive neuronal degeneration, particularly in basal ganglia and cortex, which explains the above-mentioned motor, sensory, and cognitive impairments observed throughout its progression. Amyotrophic lateral sclerosis [19, 20] is another class of progressive neurodegenerative disorder that primarily affects motor neurons in the brain and spinal cord, leading to motor impairments while sparing sensory functions in most cases of the disorder. The hallmark of amyotrophic lateral sclerosis (ALS) is motor deficits, including starting with muscle weakness, fasciculations, and spasticity. These motor symptoms progress to severe atrophy, and loss of voluntary movement, ultimately affecting critical functions such as speech, swallowing, and breathing.

3. Robotics in healthcare: An overview

The development and utilization of robotic systems in healthcare represent one of the most transformative changes in modern medicine, with robots playing a critical role in not only diagnosing a disorder but also assessing, and treating various diseases. From performing complex surgeries through teleoperation to streamlining hospital workflow, robotics in the healthcare industry has made giant strides in the recent past. Additionally, the advancements in artificial intelligence, deep learning, and biosensor technologies have enabled robots to become more responsive to human needs, expanding their application across various healthcare setups and thereby enhancing the quality, efficiency, and accessibility of medical care.

This section will provide a general overview detailing the history of robotics in healthcare. The initial applications of robotics in the neurological assessments are relatively basic and often involve complex mechanical devices and motor function examinations. These early machines/robots were more of a research tool rather than a diagnostic device, focusing on gathering data on motor and neurological disorders. The 1990s marked the beginning of utilizing robotic systems in diagnosing neurological disorders. With the increased availability of several imaging techniques such as MRI and CT scans, robots were used in tandem with these imaging techniques to assist the doctors in assessing the neurological function of the patients. In certain cases, robots were developed to guide imaging devices over certain neural regions in order to better diagnose conditions such as tremors, epilepsy, and traumatic brain injuries. Moving to the twenty-first century, it was in the early 2000s that artificial intelligence began to play a critical role in healthcare. These AI algorithms enabled robots to analyze patient's motor skills and facial expressions, thereby identifying any early biomarkers of neurological disorders. For example, several haptic devices were explored during this time to provide better insights into the patient's proprioception, which is often impaired in PD and HD. Furthermore, during this time, the field of neurorehabilitation also expanded. While prior to this, robots were mainly looked at as an analysis and diagnostic tool, with the emergence of intelligent robotic systems, the application of robotic systems in rehabilitation has been studied. Systems like MIT-Manus, a robotic arm used for rehabilitating stroke patients, allowed therapists to monitor the progress of the patients in regaining mobility, which was essential for developing patient-specific treatment plans. The 2010s saw a giant leap in robotic diagnostics for neurology as advanced systems such as brain-machine interfaces and wearable robotics gained popularity. Devices like Microsoft Kinect and several other motion sensing systems were adapted and utilized for healthcare applications, thereby enabling noninvasive assessments of gait, balance, and coordination in patients diagnosed with neurological conditions. Such systems could detect any abnormalities linked to these disorders, which in turn provide real-time data for continuous monitoring of patients. Additionally, robots were also integrated with electroencephalogram (EEG) and other brain imaging systems to assess cognitive functions, albeit in a limited manner. However, today, robotic systems are increasingly being

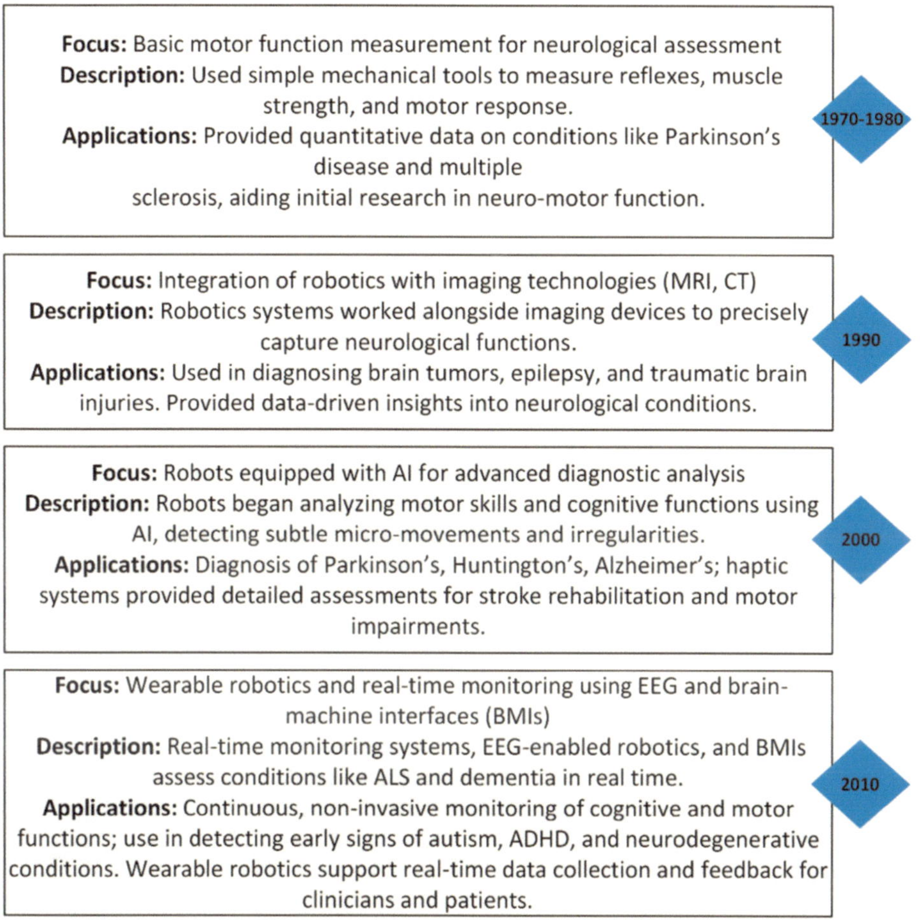

Figure 1.
Historical perceptive of robotics in healthcare.

used to provide a more objective and consistent assessment of the motor, sensory, and cognitive functions of patients diagnosed with neurological conditions. Robots can now detect micro-movements, analyze complex speech patterns, and measure eye movements, all of which contribute to the diagnosis and assessment of neurological disorders. Current advancements in robotics also focus on autonomously collecting data in a continuous stream without the need for any intervention from clinicians or technicians. With the emergence of more powerful AI systems, future robotic devices are expected to predict the onset of disease long before the symptoms appear, allowing for early intervention. In addition, due to the fact that many neurological disorders such as AD and PD have genetics as one of the etiological factors, there is ongoing research in combining robotics with genetic data to personalize assessments and treatments even further. **Figure 1** provides a historical perspective of robotics in healthcare.

4. Robotics for motor function assessment

Robotics for assessment of motion functions is a rapidly advancing field with applications in both academia and industry. These robots are not only used to assess motor function but also to assist motor functions, which are crucial for patients recovering from neurological disorders.

4.1 Key components of rehabilitation and assessment robots

Exoskeletons and prosthetics represent technologies that are aimed at enhancing the physical abilities of individuals suffering from neurological disorders and any other disabilities. Exoskeletons are wearable robotic suits that can augment human mobility by providing external force and support through actuators, sensors, and control systems. These rely on principles such as biomechanics, kinematics, and control theory to ensure precise interaction between the device and the human body. A critical component of exoskeletons is the actuators as the device may include an electric motor, pneumatic actuators, or in some cases, hydraulic systems depending on the trade-offs in terms of power, weight, and efficiency. While electric motors are common due to their ease of use and control precision, they may require high-energy batteries, which can introduce challenges in terms of power consumption and system weight. These electric motors are often controlled using pulse-width modulation (PWM) to regulate speed and torque as shown in Eq. (1).

$$\tau = K * I \tag{1}$$

where τ is the torque output of the motor, k is the motor constant, and I is the current input to the motor. Batteries, especially lithium-ion types, are the ones that are commonly used to power actuators, sensors, and control systems. However, as indicated, power consumption remains a significant challenge, especially for lower-limb exoskeletons where high forces must be generated when walking. To overcome this challenge, energy-efficient design practices such as energy recovery systems and regenerative braking are often integrated to recapture energy during certain phases of movement. In order to address the limitation caused due to the system's weight, lightweight materials, including carbon fiber and titanium, are used to reduce the overall weight of the system, although this could increase the cost of the system. On the other hand, the pneumatic and hydraulic actuators offer high power-to-weight ratios, making them suitable for lower-limb exoskeletons. For pneumatic actuators, the force is proportional to the internal pressure and cross-sectional areas of the piston (see equation). As such, by regulating the input pressure, the force exerted by the motor could be regulated as shown in Eq. (2).

$$F = P * A \tag{2}$$

where F is the force exerted by the actuator, P is the pressure within the actuator, and A is the cross-sectional area of the piston. These actuators are often controlled using closed-loop feedback systems, where real-time data from position sensors such as encoders and IMUs or force sensors is used to adjust the actuator dynamically. Recently, proprioceptive sensors were also used to measure changes in body posture and joint angles, allowing the exoskeleton to mimic natural gait patterns and support the user's movements in a seamless manner. These sensors are critical for the efficient operation of the control system, which is central to ensuring accurate interaction. Traditional control algorithms such as PID controllers [21] (see Eq. (3)) or adaptive control are often used to adjust the force or resistance provided by the actuators based on sensor feedback.

$$u(t) = K_p e(t) + K_i \int e(t)dt + K_d \frac{de(t)}{dt} \tag{3}$$

where $u(t)$ is the control input, $e(t)$ is the error (difference between the desired and actual position), and K_p, K_i, and K_d are the proportional, integral, and derivative gains, respectively. In more sophisticated systems, machine learning techniques are employed to learn optimal movement patterns and adjust to the user's behavior over time. For instance, model-predictive control (MPC) [22] can predict future movements of the user and adjust actuator output to predict and prepare for those movements. The optimization problem for MPC can be formulated as follows (see Eq. (4)).

$$\min_u \sum_{t=0}^{N} \left(||x(t) - x_{ref}(t)||_Q^2 + ||u(t)||_R^2 \right) \tag{4}$$

where $x(t)$ is the state vector at time t, $x_{ref}(t)$ is the reference trajectory, $u(t)$ is the control input (force/torque), Q and R are weighting matrixes for the states and control inputs, respectively. The objective of MPC is to minimize the error from the reference trajectory while controlling the effort expended by the actuators. The downside to using complex machine learning techniques is the lack of interpretability in machine learning models. As opposed to traditional control algorithms, the machine or deep learning models are often considered more of a "black box" as it is difficult to interpret and justify the predictions of these models based on the inputs.

On the other hand, prosthetics are artificial limbs that are designed to replace any missing body parts. Very often, modern prosthetics integrate myoelectric control systems, which utilize electromyography signals from residual muscles to interpret user intent and execute the desired movements. This may involve signal processing algorithms that filter and decode electromyography (EMG) signals to control the motion of actuators in real time. For lower-limb prosthetics such as prosthetic legs, the focus is on replicating human gait as closely as possible. This is usually achieved through a combination of mechanical joints and actuators that simulate muscle movements. The actuators and sensors may be very similar to the ones used in exoskeletons, although there may be an emphasis on making the system more energy efficient and less heavy. The upper-limb prosthetics often incorporate myoelectric control, where EMG sensors may detect muscle activity in the residual limb. These electrical signals generated by muscle contractions are then translated into control commands for the prosthetic. Advanced prosthetic hands may use multiple degrees of freedom (DOF) to allow for more dexterous movements, including fine motor tasks like gripping objects of various shapes and sizes. Due to the higher emphasis on ease of use in prosthetics over exoskeletons, wireless charging technology may be integrated into some designs, making it easier for users to keep their prosthetics powered without the need for frequent battery changes. The control systems in prosthetics may closely align with that of the exoskeleton with few minor differences.

4.2 A general overview of biomechanical modeling

In general, both exoskeletons and prosthetics leverage biomechanical modeling to mimic human joint trajectories and gait patterns. Furthermore, mathematical models such as inverse kinematics and dynamic simulations may also play a critical role in optimizing the design of these devices and ensuring smooth and natural movements. Biomechanical modeling is particularly central to the design and optimization of exoskeletons and prosthetics. These models often help researchers to simulate human motion, predict forces, and torques, and ensure that the devices can interact properly with the human body.

Kinematic models describe the motion of the body without taking into account the forces causing the motion. For exoskeletons and prosthetics, kinematics ensures that the device can replicate the natural movements performed by the users. Forward kinematics computes the position and orientation of an end-effector (in our case, a prosthetic hand or foot) based on joint angles. For a planar system with n joints, the position (x, y) of the end-effector is shown in Eqs. (5) and (6) [23].

$$x = \sum_{i=1}^{n} l_i \, cos\left(\sum_{j=1}^{i} \theta_j\right) \tag{5}$$

$$y = \sum_{i=1}^{n} l_i \, sin\left(\sum_{j=1}^{i} \theta_j\right) \tag{6}$$

where l_i is the length of the ith segment, and θ_j is the length of the jth segment. On the other hand, the inverse kinematics computes the required joint angles θ_i to achieve a desired end-effector position $(x_{desired}, y_{desired})$. This usually involves solving nonlinear equations, which are typically solved using numerical methods such as gradient descent and Jacobian-based approaches.

While kinematic models are good at estimating the position based on joint angles, they do not account for forces and torques. This is why we use dynamic models that take into account the forces and torques that produce motion. These models are crucial for controlling the actuators in exoskeletons and prosthetics to replicate human motion effectively. The motion of the limb segment can be described using Newton-Euler equations. For a segment of mass m, length l, and center of mass located at d from the joint, the torque τ at the joint is given by the following Eq. (7).

$$\tau = I\ddot{\theta} + mgdcos(\theta) + F_{ext}r \tag{7}$$

where $I = \frac{1}{3}ml^2$ is the moment of inertia of the segment, $\ddot{\theta}$ is the angular acceleration, $mgdcos(\theta)$ is the gravitational torque, and $F_{ext}r$ represents external forces (such as ground reaction force) acting at a distance r from the joint. In some cases, rather than Newton-Euler dynamics, the Lagrangian dynamics can also be used. It offers an alternative way to model limb motion, focusing on energy rather than forces. The Lagrangian L is defined as follows in Eq. (8).

$$L = T - V \tag{8}$$

where $T = \frac{1}{2}I\dot{\theta}^2$ is the kinetic energy and $V = mgdcos(\theta)$ is the potential energy. Finally, the equations of motion can be derived using the Euler–Lagrange equation as follows in Eq. (9).

$$\frac{d}{dt}\left(\frac{\partial L}{\partial \dot{\theta}}\right) - \frac{\partial L}{\partial \theta} = \tau \tag{9}$$

The Lagrangian dynamics are particularly useful for multi-joint systems. In addition to understanding and replicating human motions, it may be important to understand how muscles can generate these forces in order to replicate these forces through exoskeletons and prosthetics. The hill-type muscle model is widely used to describe muscle contraction dynamics. The total force generated by a muscle is usually considered a function of activation, length, and velocity [24] (see Eq. (10)).

$$F_{muscle} = F_{max} a f_l(l) f_v(v) \tag{10}$$

where F_{max} is the maximum isometric force, $f_l(l)$ is the force-length relationship, $f_v(v)$ is the force-velocity relationship, and a is the activation level $0 \leq a \leq 1$. Moreover, for the design of lower-limb devices, we require accurate modeling of human gait as well. This is characterized by repetitive cycles of stances and swing phases. The Ground Reaction Forces (GRF) are essential for understanding the interaction between the foot and the ground. It can be approximated using a spring-damper model as shown in Eq. (11).

$$F_{GRF} = K\delta + b\delta \tag{11}$$

where k is the spring stiffness, b is the damping coefficient, and δ is the displacement of the foot.

4.3 Advancements in robot systems for motor function assessments in academic institutions and industries

Here, the focus has been traditionally on developing and testing novel robotic systems and algorithms that can assess and improve motor function assessment. We will discuss robotic systems from two different paradigms (a) academic robots and (b) industrial robots. The robots developed in academic institutions generally focus on advancing science and are often custom-built for specific research purposes. On the other hand, industrial robots are more commonly used in practical clinical settings for rehabilitation and assessment. **Table 2** provides a comprehensive overview of robotics being used for motor assessment. Academic research focuses on making these devices more responsive to the patient's needs and more intuitive. The MIT Bio-Mechatronics lab [39] focuses on developing wearable robotic systems to augment human mobility, and the researchers work both on exoskeletons and powered prosthetics with sensors for real-time feedback. **Figure 2** shows some examples of exoskeletons developed by MIT Biomechatronics. Studies at MIT have focused on continuous neural control of these bionic limbs to restore mobility in individuals with below-knee amputation. One notable innovation from this lab is the use of EMG signals and residual limb dynamics to enable continuous neural control of bionics. This could allow for proportional control of joint trajectories in real time and also reduce the cognitive burden on the user. In terms of hardware, the lab has developed low-impedance actuators that mimic the natural dynamics of human joints. Furthermore, control systems are implemented using adaptive model-based controllers that account for user-specific gait patterns and any other environmental interactions. Studies have shown that these devices significantly reduce metabolic costs during locomotion, improve balance, and provide enhanced proprioceptive feedback. This, in turn, contributes to a more natural and comfortable user experience.

The harmony exoskeleton from UC Berkeley [40] is another such robotic system that has garnered attention in recent years. It is designed for lower-body rehabilitation, particularly for stroke patients, and allows researchers to study the gait movements of the limbs affected due to disorders. One notable work at Berkeley Robotics is the Berkeley Lower Extremity Exoskeleton (BLEEX) (shown in **Figure 3**), which is

Type	Robot category	Robot	Features
Academia	Exoskeletons and prosthetics	MIT Biomechatronics Lab [25]	Adaptive-powered prosthetics equipped with sensors for real-time feedback.
		Berkeley Robotics BLEEX [26]	Load-sharing, lower-body exoskeleton with energy-efficient actuators and stability tracking.
	Rehabilitation robots	Vanderbilt Prosthetic Legs [27]	Powered joints for adaptive motion, and sensors for gait tracking, designed to support natural movement patterns.
	Neuro-robotics and BMIs	Brown University BMI Prosthetic	Brain-machine interface technology interprets neural signals to control robotic limbs and supports adaptive motor learning.
Industry	Exoskeletons and prosthetics	ReWalk Exoskeleton [28, 29]	FDA-approved, customizable to user height and weight, provides real-time gait tracking, programmable for rehabilitation routines.
		EksoGT [30, 31]	Adjustable support for different levels of assistance, advanced motorized joints, and data logging for therapy monitoring.
	Rehabilitation robots	Kinova JACO Arm [32]	Lightweight, multi-degree-of-freedom robotic arm, precision control for tasks requiring fine motor skills, and customizable grip strength.
		Hocoma Armeo [33]	Sensor-rich robotic arm exoskeleton, tracks hand and arm movements, and adjustable resistance for therapy progression.
		KINARM [34]	Robot manipulator and exoskeleton equipped with a virtual reality platform featuring high-resolution motion capture and customizable task environments for neurological and biomechanical research.
	Motion Capture and Kinematic Analysis	IntelliRehab Motion Platform [35]	Wearable sensors for body motion analysis, automated gait and posture data collection, and software for data visualization and tracking improvements.
		GaitUp Sensors [36]	Wearable motion sensors with accelerometers, gyroscopes, real-time gait, and balance tracking. Portable for use outside clinical settings.
	Neuro-robotics and BMIs	Myomo MyoPro Orthotics [37]	AI-driven robotic arm orthosis detects EMG signals to assist hand and arm movements, and personalized learning algorithms for user adaptation.
		Bionik Labs InMotion Robot [38]	Adaptive AI that adjusts to a patient's progress, precise joint movement tracking, and automated data collection for clinical insights.

Table 2.
A comprehensive list of robotic systems for motor assessments.

designed to help soldiers and emergency personnel carry heavy loads with minimal effort across diverse terrains. The system encompasses powered legs, a power unit, and a frame for carrying equipment, enabling users to offload loads without impacting their agility. Unlike the earlier exoskeleton, this technology is aimed at assisting

Figure 2.
Neurally controlled prosthesis by MIT biomechatronics [25].

Figure 3.
Berkeley lower extremity exoskeleton [26].

soldiers and other emergency personnel to carry heavy loads. BLEEX incorporates a hybrid power source that provides hydraulic power for locomotion and electrical power for on-board computers. Similar to several other exoskeletons, this system also relies on sensor data from the exoskeleton to estimate wearer movements without the requirements of direct measurements from the human. Finally, BLEEX also utilizes

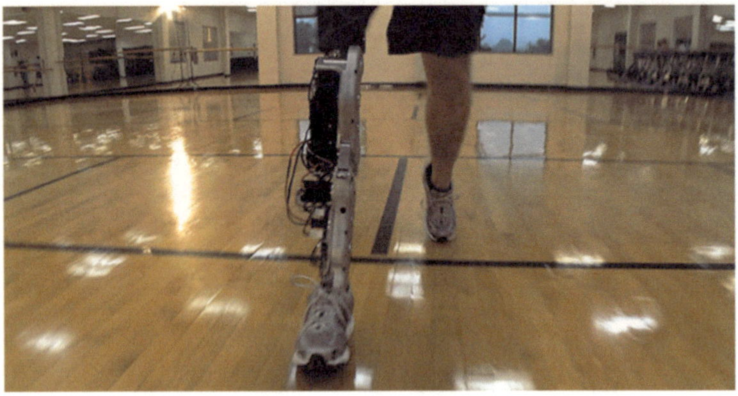

Figure 4.
Vanderbilt prosthetics and robotics [27].

high-speed body LAN with a synchronous ring network to manage communication between electronic modules, sensors, and actuators present in the system.

Vanderbilt University's research into lower-limb prosthetics (shown in **Figure 4**) offers improved stability and mobility for amputees. They incorporate adaptive control algorithms to mimic natural gait patterns. Many researchers have also expended a considerable amount of time and effort into neuro-robotics, which involves the study of integrating robotics with neuroscience to understand how the brain controls motor functions. The work led by Vanderbilt University's Center for Intelligent Mechatronics includes the development of the first robotic leg with powered knee and ankle joints. The design is equipped with a neural interface and it allows for the control of the robotic system by thought. The research team has highlighted that powered prosthetics help users to walk faster with less hip effort and energy use as opposed to passive limbs.

In addition to the development of lower-limb prosthetics, the team has also investigated the impact of these prosthetics on the users. This study [41] explores the biomechanical effects of adding an articulating toe joint to passive foot prosthetics for incline and decline walking, which indicates that there are relatively small effects on joint kinematics and kinetics during sloped walking. Additionally, this team [42] also investigated the needs and requirements that need to be considered when developing a bimodal foot prosthesis for walking and running. These requirements have been considered when developing these prosthetics and can be used/built upon by future researchers focused on developing prosthetics.

Brown University has focused on developing Brain-Machine Interfaces (BMI), which allow users to control robotic limbs using brain activity. These robotic systems are assessed for both motor function restoration and to study neuroplasticity in the human brain. The BrainGate research program has developed brain-computer interface technologies to assist individuals with neurological conditions, injuries or limb loss regain communication, mobility, and independence. By implanting microelectrode arrays in the brain, the developed technology decodes neural signals associated with movement intentions to control external devices. As per the results, the proposed system enabled people with paralysis from conditions like spinal cord injuries, brainstem strokes, and ALS to operate computer cursors, advanced prosthetics, communication aids, and in some cases even their paralyzed limbs [43].

Deo et al. [44] have explored the possibility of utilizing neural network decoders to restore multi-effector motion such as bimanual movement for people with paralysis. The researcher developed a recurrent neural network model for such movements by altering the temporal structure of training data. It demonstrated that a person with paralysis could control two computer cursors using AI techniques such as RNN decoders. This use of RNNs to enable bimanual control is a novel approach and can be used to address several challenges pertaining to brain-computer interfaces especially when performing complicated movements and tasks.

Several other research institutes across multiple universities have also worked toward the development of cutting-edge robotic systems controlled through neural signals [45].

Apart from academic research, several industrial products focused on the rehabilitation of patients suffering from neurological disorders have also been developed over the years. The ReWalk exoskeleton [28, 29] (shown in **Figure 5**) is one such product developed by Lifeward. The ReWalk Exoskeleton is a wearable robotic system designed to assist individuals with lower-limb paralysis regain mobility and independence. The exoskeleton achieves this through a combination of advanced automation, real-time motion sensing, and an intuitive user interface. The device is powered by electric motors at the hip and knee joints and through this, it generates the torque needed for locomotion. The tilt sensors detect shifts in the user's upper body to coordinate steps and improve device control. Discussing the structural design, the system's modular adjustability ensures compatibility with a range of body sizes, making it accessible to individuals with varying levels of mobility impairment.

Another noteworthy contribution of this device is the development of a wrist-mounted remote control that enables users to switch between operational modes

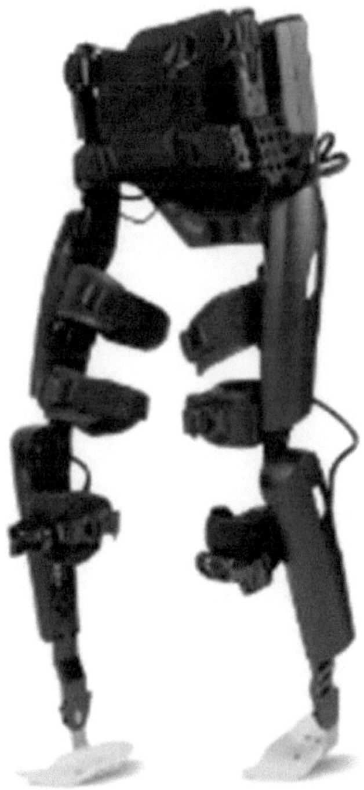

Figure 5.
Rewalk [46].

(standing, walking, sitting, or climbing stairs). Future work pertaining to this device may focus on battery life, device miniaturization, and system cost. Developed primarily for people with spinal cord injuries, it provides powered hip and knee motion, allowing users to stand, walk, and even climb stairs using crutches and walkers for balance. Apart from spinal cord injuries, this exoskeleton may also be useful for other neurological conditions such as Parkinson's disease which impacts the mobility in patients. The exoskeleton is controlled by subtle shifts in body weight and movement, and this is combined with a computer-based control system and sensors to enable a more natural waking gait.

EksoGT [30, 31] (shown in **Figure 6**) is another advanced wearable exoskeleton designed primarily for rehabilitation and assessment purposes, specifically assisting patients recovering from a stroke, spinal cord injuries, and any other forms of lower-limb paralysis. Manufactured by Ekso Bionics, the device uses personalized software for therapies and allows patients to adjust the level of support and step initiation depending on their needs. The robotic exoskeleton relies on a variety of advanced technologies for its precise and adaptive assistance. For sensing, it encompasses force sensors for measuring ground reaction forces, IMUs to track the user's movement, and position encoders to monitor joint angles. Apart from this, it also includes some actuation systems coupled with closed-loop control to ensure adaptive joint movements. Recent studies have also focused on the development of a vibrotactile feedback system to improve balance rehabilitation in the EksoGT exoskeleton. This study discusses methods to utilize sensors to detect users' balance and then provide vibrotactile cues to help them adjust their posture and maintain stability. Many earlier works [48] have discussed the effectiveness of using sensory cues to improve the motor performance of patients diagnosed with neurological disorders. A position-based sensory feedback strategy was implemented based on the difference between the desired and

Figure 6.
EksoGT [47].

current position. The sensory feedback from these auxiliary sensors/devices is aimed to mimic the feedback from the therapist. This robot represents a significant leap in assistive and rehabilitative technology and one of the novel aspects of this robot is its suitability for patients with varying levels of ability. In many ways, both Rewalk Exoskeleton and EksoGT serve a similar purpose of helping individuals with disabilities. The major difference is that while ReWalk is typically used both in clinical settings and at home, the EksoGT focuses more on rehabilitation with advanced features like dynamic support. **Figure 6** shows an image of EksoGT.

Meanwhile, apart from exoskeletons, robotic arms like Kinova JACO Arm [32] and Hocoma Armeo [33] provide crucial assistance to individuals with upper-limb paralysis, weakness, or disabilities. The JACO arm is a highly adaptable, lightweight robot arm that can be controlled by the user's movements or through external switches. This enables individuals to perform day-to-day tasks like eating, using a computer, or grabbing objects. With several neurological disorders such as PD and HD, impacting patient's ability to perform day-to-day activities, these robots with real-time feedback and adjustable assistance could greatly assist the patients in performing their day-to-day activities. Furthermore, it also encourages patients to engage in repetitive movements that are vital for neural recovery.

Moving away from upper-limb rehabilitation, devices such as IntelliRehab Motion platform [35], Gaitup Sensors [36], and MyoPro Orthotics [37] focus on enhancing movement analysis and rehabilitation at a more personalized level. The IntelliRehab motion platform uses sensors to assess a patient's gait and balance. Similarly, Gaitup Sensors monitor detailed movement data, enabling clinicians to track progress and refine therapy plans. Lastly, MyoPro Orthotics uses electromyography (EMG) signals to enable individuals with weak upper limbs to regain some level of independence. Another key robot system that has garnered a lot of attention in recent years is the KINARM [34], which is an advanced robot system designed for clinical assessment and rehabilitation of the upper limb in patients with neurological disorders. KINARM consists of a robot system and a virtual reality (VR) system that can track and guide a patient's upper-limb movement. The system offers a wide range of tests to measure a patient's ability to perform specific tasks like reaching, tracking, and object manipulation. Furthermore, it also offers the technicians and engineers the opportunity to develop their custom test experiments based on the requirements, which is extremely critical to explore newer symptoms that may have remained hidden so far. Several clinical studies [49, 50] have utilized these devices to study the motor and non-motor functions that are impacted due to various disorders. Furthermore, the in-depth data provided by these robots could assist in better understanding the disorders and also help develop a more targeted and personalized treatment.

In conclusion, the development of robotic devices for the assessment and rehabilitation of patients with neurological disorders has made giant strides over the last decade. These developments have offered promising solutions not only for improving mobility and independence but also for a better understanding of the pathology of various neurological disorders. Exoskeletons such as ReWalk and EksoGT provide great advancements in lower-limb rehabilitation, assisting individuals with spinal cord injuries. Meanwhile, robot arms such as Kinova JACO Arm and Hocoma Armeo provide critical support for supper-limb rehabilitation. Finally, devices such as IntelliRehab, Motion Platform, gaitUpSensors, MyoPro Orthotics, and the KINARM further enhance personalized treatment plans and offer detailed movement analysis and tracking. These robot and sensor systems have also integrated emerging technologies such as VR for more effective rehabilitation strategies.

5. Robotics for cognitive assessment

Numerous robot systems have been developed for assessing motor capabilities in patients with neurological disorders and are also capable of evaluating sensory and cognitive functions, including sensorimotor integration, sensorimotor control, and non-motor decision-making. Several robots that have been discussed earlier for motor assessment may also be utilized for cognitive assessment. In this section, we will be discussing specific robot systems aimed at cognitive and non-motor assessment. In this section, again, we will be discussing robots from academic and industrial research.

Robots have emerged as a valuable tool for cognitive assessment owing to their precision, adaptability, and ability to provide consistent and repeatable testing environments. Traditional cognitive evaluations often rely on human administration, which can introduce variability leading to results that may not be consistent. Robots, on the other hand, can standardize the process and offer greater consistency, which is crucial for accurate assessments. More importantly, robots can also create engaging and interactive testing environments, which are critical for cognitive assessment. For example, robots designed with friendly, interactive, and approachable features can reduce anxiety or resistance during assessments. Through integration with certain interactive technologies such as virtual reality, these robots can also employ gamified tasks that test memory, attention, and problem-solving skills, all of which are found to be impacted due to neurological disorders. Finally, the integration of artificial intelligence with robot systems allows for adaptive testing, where the difficulty and nature of cognitive tasks can be dynamically adjusted based on the performance of the user. This personalized approach ensures that the assessment is neither too easy nor too challenging, providing a more accurate evaluation of an individual's cognitive abilities. These robots can also analyze data they collect in real time, and identify any patterns that can point to neurological disorders, which makes them not only efficient for assessment and monitoring but also invaluable for diagnosis.

5.1 Human-robot interaction

In the context of cognitive assessment, the topic of human-robot interaction is of vital importance. It focuses on creating an accurate and engaging interaction that can facilitate the evaluation of cognitive abilities such as memory, attention, problem-solving, and executive function. The key components in human-robot interactions for cognitive assessment include communication systems, sensor technology, adaptive testing algorithms, and real-time feedback, and interaction.

5.1.1 Communication systems

Robots aimed at cognitive assessment often rely on multimodal interaction to ensure robust data collection and user engagement. For speech recognition and processing, robots may use natural language processing to interpret verbal instructions, responses, and cues. These speech systems must be capable of handling diverse accents, speech patterns, and any other potential impairments. On the other hand, for nonverbal communication, robots equipped with cameras and motion sensors can capture nonverbal behaviors such as facial expressions, eye gaze, and hand movements. For visual data, advanced image processing algorithms can be used to analyze these cues to identify subtle signs of cognitive impairment.

5.1.2 Sensor technology and data capture

Human-robot interaction for cognitive assessment needs accurate data collection from multiple sensors, including visual, auditory, and physiological sensors. For visual data, high-resolution RGB and depth cameras can be used for tracking facial expressions, eye movements, and physical responses. Further, microphones are often used to capture audio data for speech analysis and tone modulation. Depending on the target users, noise reduction and filtering algorithms may be needed to ensure the accuracy and reliability of the captured audio signals. Finally, a few systems may also possess physiological sensors, including heart rate monitors, skin conductance sensors, or EEG headsets for real-time monitoring of stress, attention, or mental effort.

5.1.3 Adaptive testing algorithms

Adaptive testing algorithms are critical to tailor test difficulty and content based on user performance, which enables clinicians to provide a patient-specific treatment approach. The widely proposed method is a form of Bayesian optimization algorithm or reinforcement learning which can be employed to adjust the sequence and difficulty of task dynamically. The real-time behavior or performance of the participants needs to be fed into a decision-making model to refine task administration. In some cases, robots can also use task-switching algorithms to assess executive function by presenting alternating activities, which requires rapid adaptability in task generation and response interpretation.

5.1.4 Real-time feedback and interaction

Real-time responsiveness is crucial for maintaining user engagement. In many cases, the robots may provide visual, haptic, vibrotactile, and auditory feedback to the users. Further, robots may use dialog systems to provide feedback such as confirming answers or encouraging the participant using NLP frameworks like Rasa or Google Dialogflow which are often useful for conversational management. Another important aspect to consider is latency optimization. Real-time interaction demands low-latency computation, often achieved through edge computing for local processing of sensor data. Most robots use edge computing as opposed to cloud servers due to privacy concerns and reduced latency.

5.2 Humanoid robots

Humanoid robots, with their human-like appearance and interactive capabilities, are well-suited for conducting cognitive assessments. The key components discussed in human-robot interaction apply to humanoid robots. The humanoid robot's ability to integrate communication systems, adaptive algorithms, and feedback mechanisms allows it to evaluate cognitive abilities like memory, attention, executive function, and problem-solving skills. In the next section, we will discuss a few of these humanoid robots that have been used for cognitive assessments.

5.3 Advancements in robot systems for cognitive function assessments in academic institutions and industries

We will begin with discussing robots from academia. **Table 3** provides a comprehensive overview of robots being used for cognitive assessment. The iCub robot [51] is

Type	Robot	Features
Academia	iCub Robot [51]	Designed for cognitive and sensory research with a focus on tasks like memory, problem-solving, and sensory-motor integration
	KASPAR Robot [52]	Used for social and cognitive assessments primarily for children with autism
Industry	Nao Robot [53, 54]	Humanoid robots designed for interaction and engagement are widely used in therapy and cognitive assessment
	Paro Robot [55]	The therapeutic robot is designed as a robotic seal to help assess and improve emotional engagement and cognitive responses
	iPal Robot [56]	Social robots aimed at assisting children with autism in social and cognitive development

Table 3.
A comprehensive list of robotic systems for cognitive assessment.

a humanoid robot developed to research cognitive development, human-robot interaction, and sensory-motor integration. It was developed by Instituto Italiano di Tecnologia (IIT) and is a part of the RobotCub project, which aims to advance the understanding of human cognition and how robot systems can learn and develop through interacting with their surroundings. The robot has 53 degrees of freedom, which allows for highly articulated movements such as reaching, grasping, and performing several other motor tasks. One of the novel features of this robot is the fact that the robot's cognitive functions are developed based on human sensorimotor coordination and mapping. Building on several behavioral and fMRI studies, the researchers were able to develop a controller system for several tasks, including spinal behaviors, eye movements, and reaching and grasping movements. This opens new doors for investigating sensorimotor integration and control in patients with neurological disorders. Very little is currently known about these sensorimotor functions and the related deficits caused due to neurological disorders. Therefore, a tool such as this that takes into account the complex sensorimotor models that may be employed by the Central Nervous System (CNS) may be very useful in studying these deficits. The iCub robot may also be considered a powerful tool for studying how humans learn, especially through embodied cognition—the idea that cognition is shaped by interactions with the physical world. It is extensively used in studies related to autism spectrum disorder, cognitive development, and human-robot collaboration.

Another humanoid robot developed at the University of Hertfordshire is the KASPAR robot [52], which can assist in cognitive and social development research. KASPAR is designed to engage in social interaction with children in a way that is supportive, thereby providing a platform for researchers to explore social and emotional learning in children with developmental disorders. Its primary purpose is to help children with autism spectrum disorder practice and develop social skills in a controlled and safe environment. The robot can engage in various activities, including interactive games, role-playing, and even emotion recognition tasks, where it mimics human behaviors and expressions. This makes the KASPAR robot highly valuable in therapeutic settings, where children can learn to recognize emotions, respond to social cues, and improve communication skills without any pressure from human interactions. Unlike general-purpose humanoid robots, KASPAR is specifically designed to act as a social mediator, providing a controlled environment to assist individuals learn and practice social and communication skills.

Moving to industrial robots, the Nao Robot [53, 54], developed by Softbank Robotics is a humanoid robot widely used in healthcare to explore human-robot interaction and cognitive development. Noa is equipped with advanced sensors, actuators, and AI capabilities, making it an ideal platform for cognitive assessment, social interaction, and therapeutic applications. Nao Robot excels at engaging users through several interactive activities that evaluate people's memory, attention, problem-solving, and decision-making. One notable example of the robot's capabilities is that the Nao Robot can guide users through memory games, puzzles, or storytelling tasks, which provides immediate feedback and assists in adjusting task difficulty based on the user's performance. These capabilities make it very valuable for studying conditions such as autism spectrum disorder (ASD), attention deficit hyperactivity disorder (ADHD), and early signs of dementia. Extensive reviews have been conducted to evaluate the performance of users in multiple categories including social interactions, assisted teaching, autism, cognitive impairments, and intervention. Another therapeutic robot designed to provide emotional support and stimulation for patients with cognitive impairments is the Paro Robot [55]. It is designed after a baby harp seal and it is equipped with soft fur and life-like movements that can encourage tactile interaction and comfort. The Paro Robot is widely used for elderly patients diagnosed with Alzheimer's disease and related dementias. A notable contribution is the Paro Robot's ability to provide companionship to patients, thereby assisting in alleviating symptoms like anxiety. Discussing robots focusing on caregiving purposes, the iPal robot [56] is a social and interactive humanoid robot developed by AvatarMind. It has gained attention for its potential applications in healthcare, especially for engaging with patients suffering from cognitive or developmental challenges. The robot is equipped with advanced AI capabilities, making the robot adaptable for use with children, elderly individuals, and more importantly, patients with neurodegenerative disorders. For patients diagnosed with Alzheimer's or PD, iPal can help stimulate cognitive and motor functions through interactive games and guided physical activities. Furthermore, equipped with sensors, the robot can monitor physical activities and collect vital patient data on user engagement. Finally, the robot is designed to not only identify and respond to emotions, thereby fostering companionship and reducing any feelings of isolation, thereby assisting in the management of anxiety.

In summary, recent advancements in robotic systems have shown great promise in the assessment of cognitive systems by offering innovative solutions to address the limitations of traditional methods. Academic robots such as iCub and KASPAR have advanced our understanding of cognitive development and sensory-motor integration. On the industrial side, robots like Nao, Paro, and iPal have demonstrated their efficiency in therapeutic settings, helping individuals improve their cognitive abilities and manage emotional well-being. These advancements highlight the potential of robotics in the management of neurodegenerative disorders by providing tools that are not only scalable and interactive by also capable of delivering profound insights into patient health and behavior.

6. Quantitative metrics and data-driven insights

Quantitative metrics and data-driven insights are extremely important for objectively assessing motor, sensory, and cognitive functions, especially in robotics-assisted interventions and assessments. These metrics enable clinicians and researchers to not only evaluate patient performance but also track the progress of any disorders and

Type	Metrics	Use cases
Motor function assessment metrics		
Kinematic data	Range of motion (ROM), velocity, acceleration, jerk (smoothness), joint angles	Analyzing arm trajectory in stroke rehabilitation to monitor recovery
Kinetic data	Force, torque, grip strength	Evaluating muscle strength in exoskeleton-assisted therapy
Temporal metrics	Reaction time, time-to-completion, movement onset, latency	Measuring the delay in initiating movements for ALS patients
Error metrics	Deviation from a predefined trajectory, task completion accuracy, error rates	Assessing hand-eye coordination using robotic systems in fine motor tasks
Electromyography (EMG) data	Muscle activation patterns, amplitude, frequency	Monitoring muscle rehabilitation progress in spinal cord injury patients
Cognitive function assessment metrics		
Memory metrics	Recall accuracy, response latency, working memory, capacity	Testing cognitive recall using interactive robot quizzes for dementia patients
Attention and focus	Fixation duration (eye tracking), Task adherence, Error frequency	Monitoring attention span in children with ADHD using robots like Nao
Decision-making metrics	Accuracy, choice latency, risk assessment tendencies	Using robots to evaluate decision-making skills in simulated problem-solving tasks
Behavioral metrics	Social cue recognition, turn-taking, emotional response latency	Evaluating emotion recognition in children with autism through robot interaction
Neurophysiological metrics	EEG patterns, ERP (event-related potentials), and BCI-based focus levels	Monitoring attention or cognitive load in real-time during robot-assisted learning

Table 4.
Quantitative metrics for motor and cognitive assessments.

design personalized therapies. In motor assessments, metrics such as range of motion, velocity, reaction time, and force offer great insights into muscle strength, coordination, and movement efficiency, while kinematic and kinetic data help track recovery in rehabilitation. On the other hand, cognitive assessment leverages metrics such as recall accuracy, response latency, attention span, and decision-making accuracy to evaluate memory, focus, and executive function. Furthermore, advanced tools such as electromyography (EMG), eye tracking, and brain-computer interfaces (BCIs) enable precise measurements of neuromuscular activity, visual attention, and neural engagement. These data points allow for longitudinal trend analysis, benchmarking against healthy baselines, and the personalization of interventions. Moreover, advanced machine learning algorithms using these data points can detect anomalies, predict risks, and adapt therapies in real time, making robotic systems indispensable for assessing and improving motor and cognitive performance. **Table 4** discusses several metrics utilized for motor and cognitive assessments.

7. Discussion

In this chapter, we discussed the basics of robotic systems used for motor and cognitive assessment and also explored a few novel robot systems developed by academic institutions and industries for the rehabilitation and assessment of neurological disorders. In the realm of rehabilitation and assistive technologies, the design, features, and application of robotic systems vary significantly, thereby reflecting diverse user needs and clinical goals. Different categories of robot technologies such as neurally controlled systems, exoskeletons, prosthetics, and brain-machine interfaces have been detailed in this chapter. Each of these robot systems comes with its own strengths and limitations, which makes it suitable for certain tasks and not so much for others. These limitations should be explored to better understand the future work needed in this field to enhance the rehabilitation and assessment capabilities in the clinical world.

In recent years, neurally controlled systems have been a center of attention due to their applications in rehabilitation. The development of neurally controlled prosthetics and exoskeletons showcases the synergy between the field of robotics and neuroscience. The work by the MIT Biomechatronics team on continuous neural control through EMG signals and residual limb dynamics highlights the potential for the integration of robotic systems with the human nervous system. These systems may not only reduce the cognitive burden on the users but also ensure that movement is more natural, which is crucial for user acceptance and long-term utility. Furthermore, the low-impedance actuators developed by the same lab mimic the natural dynamics of the human joints and may assist with optimal functionality. Similarly, Brown University's BrainGate research also demonstrates the transformative potential of BMI technologies. By enabling individuals with paralysis to control external devices through neural signals, the system opens the possibilities for restoring independence in individuals with several motor impairments. Both the BrainGate and robots from MIT Biomechatronics integrate advanced interfaces to enable intuitive control. However, they differ in the invasiveness and the target populations. While BrainGate's direct neural interface provides control for individuals with paralysis, BMI often involves implantable technologies, which makes it more invasive. However, MIT Biomechatronics systems use EMG signals, offering a noninvasive alternative aimed at amputees. It also needs to be noted that several studies [24, 57] have discussed the extensive need to properly filter the EMG signals, without which they might not be accurate or reliable. Therefore, both these systems have their limitations. While one may be invasive, the other system may need to ensure that the received EMG signals are appropriately processed to ensure accuracy and reliability. Moving forward, the research may focus on minimally invasive brain implants that can complement the EMG data to balance precision with comfort.

Apart from neural-controlled devices, another key point of discussion is the adaptive control systems. The adaptiveness of the exoskeletons and prosthetics is of paramount importance. The adaptive control systems developed by institutions such as Vanderbilt University highlight the importance of tailoring robotic systems to individual user needs. The use of powered prosthetics with adaptive control algorithms allows for a more natural gait movement and also reduces the metabolic sot of locomotion. However, these systems face challenges related to power consumption, durability, and integration with diverse physiological conditions, which need to be addressed to ensure widespread adoption.

Moving away from lower-limb systems, robot systems such as Kinova JACO arm and Hocoma Armeo highlight the role of robotics in upper-limb rehabilitation and assistance. These devices not only restore functionality but also encourage repetitive movements, which are considered vital for neural recovery. Furthermore, KINARM's integration of virtual reality with robotic devices provides a unique platform for clinicians to study and track patient progress in a highly interactive and controlled environment. While both these systems (Kinova JACO arm and Hocoma Armeo) facilitate repetitive exercises to improve motor function, their focus areas are different. Kinova JACO emphasizes assistive use, thereby providing continuous support for daily activities through adaptable and precise control mechanisms. In contrast, the Hocoma Armeo focuses more on clinical rehabilitation that incorporates a virtual reality environment to engage patients and thereby encourage quicker recovery.

With the emergence of high-precision healthcare systems, including medical robotics and AI, the focus on personalized treatments has increased drastically. Several studies [58, 59] have also shown how a patient-specific treatment may lead to better outcomes for patients as opposed to generic treatments. Platforms like IntelliRehab motion and Gaitup Sensors represent a paradigm shift in therapy. By utilizing sensor data and advanced analytics, these systems allow for real-time monitoring of patient progress and provide clinicians with actionable insights to refine treatment plans. Similarly, the use of EMG signals in MyoPro Orthotics further highlights how data-driven approaches can assist individuals with weak upper limbs to regain independence. While this is promising, much more can be achieved by leveraging the lower-limb and upper-limb data from assessment and utilizing AI for more in-depth analysis of the patient data, which enables clinicians to provide a more patient-specific treatment based on symptoms. Many subtle differences in a patient's condition which may not be noticeable by traditional assessment could be captured through these high-precision robot systems, which can be extremely beneficial in characterizing the patient's symptoms and tailoring the treatments based on them.

Feedback received from the robot system is crucial in assistive and rehabilitation technologies as it would enable the patients to improve their motor and non-motor performance over time. Different robot systems employ varying methods to provide sensory feedback. Ideally, the type of feedback needs to be tailored to the target user groups. For instance, patients suffering from visual dysfunctions should not be provided only with visual cues but should rather be complemented with vibrotactile or haptic feedback. Similarly, depending on the task at hand, the modality and type of feedback may also need to be altered. Studies [60] have shown that certain sensory modalities may take precedence over others depending on the task at hand. Exoskeletons such as EksoGT and robotic arms such as Kinova and JACO provide users with tactile and force-based feedback to replicate neural interactions. In EksoGT, haptic feedback is integrated to assist with balance training by signaling shifts in weight distribution, thereby assisting users regain mobility skills. Similarly, Kinova JACO uses force sensors to ensure smooth and precise movements during assistive tasks. This type of feedback greatly helps with motor rehabilitation as it reinforces neural pathways associated with tactile sensations and movement control. Systems like MyoPro Orthotics use vibrotactile feedback to signal muscle activation levels, thereby helping patients modulate their efforts during therapy. These cues are especially effective because they provide non-intrusive guidance, enabling the users to focus on their tasks while receiving real-time feedback. A less commonly used or emphasized feedback is auditory feedback. Studies [49, 50] have provided auditory feedback through the KINARM robot to assess sensorimotor control and integration in Parkinson's patients. Finally, the most widely used sensory feedback is the visual feedback. Systems such as Hocoma Armeo and IntelliRehab Motion, which incorporate virtual reality provide visual feedback to the users. Hocoma Armeo uses immersive virtual reality environments to engage patients, allowing them to visualize their movements in real time and track their progress. IntelliRehab Motion provides a real-time visual representation of joint angles and movement trajectories, enabling patients and therapists to monitor performance and make adjustments. Despite the several advancements in providing sensory feedback, the most critical limitation is that currently, we do not optimize the sensory feedback to suit individual needs. Additionally, providing sensory feedback for individuals with severe sensory deficits poses significant challenges as these users may not be able to appropriately perceive and interpret their feedback. Finally, it may be beneficial to provide feedback through multiple modalities rather than a single modality. Sensorimotor integration usually gathers inputs from multiple modalities to understand fully the state of the world around the user. Different modalities may be suitable to effectively convey different information. Therefore, providing multimodal feedback allows the users to get a more holistic understanding of the environment around them.

Moving to cognitive assessments, robots such as iCub, KASPAR, Nao, Paro, and iPal have great applications in therapeutic interventions and caregiving. Robots such as iCub and KASPAR focus on cognitive and social development. iCub's ability to replicate human-like sensorimotor coordination makes it a one-of-a-kind tool for studying complex neural functions and their deficits in conditions such as Parkinson's disease. While iCub uses specific sensorimotor models to replicate human-like movement and cognitive function, it needs to be noted that there is no single sensorimotor model that is clinically validated or can fully comprehend or explain how the central nervous system (CNS) performs several diverse and complex functions. Studies [49, 50, 60] have shown the CNS uses a more adaptive sensorimotor strategy based on reinforcement learning wherein the mathematical or computational models are utilized to achieve a specific goal will differ based on several factors including the task at hand, internal, and external states. Similarly, the models for sensorimotor integration are also found to be adaptive wherein the CNS may determine a specific strategy based on several internal and external factors. While the integration of sensorimotor principles in the iCub robot is very promising, more work is needed to better understand the human sensorimotor principles, which would also enable us to more closely mimic human cognitive functions. While the iCub robot focuses on studying complex neural functions, KASAPR, on the other hand, is designed for social interaction, particularly for children with autism spectrum disorder, providing a safe and controlled environment to practice social skills. Finally, industrial robots such as Nao and Paro emphasize therapeutic applications, with Nao engaging users with interactive cognitive tasks and Paro offering emotional comfort to elderly patients with dementia. Despite these advancements, these robots come with their unique strengths and limitations. While iCub excels in replicating human sensorimotor functions, the model utilized in the robot may not fully account for the wide range of sensorimotor models adopted by the CNS. KASPAR is highly effective for targeted social interventions but very little is known about the robot's applicability for studying neural functions and physical rehabilitation. Finally, across all the robots, a common limitation is their reliance on pre-programmed behaviors and limited adaptability to highly dynamic or unpredictable environments. Additionally, the cost of these robots remains a significant barrier to widespread adoption, particularly in resource-constrained settings.

8. Challenges and limitations

In the last few decades, several advancements have been made in the development of cutting-edge robot systems for the rehabilitation and assessment of neurological disorders. However, there still exist several limitations in the existing robot systems that have been developed.

Discussing the technological challenges, neurally controlled systems such as BrainGate and the robot from MIT Biomechatronics have demonstrated significant potential for restoring motor function. However, their effectiveness may be limited by key challenges such as the invasiveness of BrainGate's direct neural interface, which may require surgical implantation, or ensuring the signal accuracy and reliability in EMG signals used by MIT Biomechatronics. Future developments in signal processing and minimally invasive technologies could mitigate these issues, thereby balancing precision and comfort for users. Another key challenge is the adaptability of robotic systems, which is critical for tailoring their functionality to individual user needs. While prosthetics with adaptive control algorithms, like those developed at Vanderbilt University, reduce metabolic costs and improve gait, their integration into diverse physiological conditions still remains a challenge. This is because these systems often require significant power consumption, limiting their long-term usability. Sensory feedback is another pivotal aspect that is critical for effective assistive and rehabilitation technologies. Yet current systems exhibit several limitations. Currently, feedback is not optimized to the individual's needs, which can reduce its effectiveness. For example, systems that rely solely on visual feedback may not be suitable for users with visual impairments and single-modality feedback may fail to provide a comprehensive understanding of the environment. Studies [49, 50] have developed and discussed multimodal feedback systems. However, more sophisticated and intelligent multimodal systems that can adapt to the user needs are required for improved effectiveness. Additionally, robots such as iCub, KASPAR, Nao, Paro, and iPal have demonstrated diverse applications from studying neural functions to providing social and emotional support. However, these robots face challenges in adaptability and cost. The iCub's sensorimotor models are promising for studying human sensorimotor coordination and are noteworthy. However, due to our lack of understanding of human sensorimotor models, currently, we do not have a robot system that can fully replicate the complex CNS's adaptive and dynamic sensorimotor strategies. Other robots may rely on pre-programmed behaviors, reducing their effectiveness in dynamic or unpredictable environments. Moreover, it may be necessary to extract features and meaningful insights from high-volume multimodal data generated by robotic systems. Recent advances in machine learning have provided us with several tools to handle such a scenario. However, it must be noted that these advanced algorithms and robust fusion techniques require a significant computational resource, which could also increase the overall cost of the assessment. Finally, due to the complexity of the disorders, robotic systems must adapt to the unique needs and capabilities of individual patients. This may require sophisticated AI and machine learning models capable of personalization.

Moving to clinical challenges, the complexity of neurodegenerative symptoms is extremely high. Neurodegenerative disorders often present multifaceted motor, cognitive, and sensory impairments as discussed in the earlier sections of the chapter. Therefore, it may be very difficult for a robotic system to comprehensively assess such a complex system without human intervention. Further, there is a lack of standardized protocols to validate the reliability, accuracy, and clinical relevance of robotic assessments when compared with the traditional methods. Unlike traditional methods that have undergone extensive validation over decades, robotic assessments often lack robust, comparative studies to establish their reliability and accuracy. Therefore, it may be difficult to ensure that these robotic assessments are consistent across different patient populations and clinical settings. Moreover, introducing and integrating robotic systems into well-established clinical workflows can be disruptive to the point that it can be logistically challenging to the clinicians. Clinical staff may need extensive training to operate and interpret robotic systems effectively, which can strain the already limited healthcare resources. Finally and most importantly, ethical considerations are a critical barrier to the widespread use of robotics in neurodegenerative care. It is vital that patient privacy and data security are ensured, especially as robotic systems often collect large volumes of extremely sensitive data. In addition to privacy concerns, the use of AI-driven analysis may also introduce challenges related to transparency and interpretability, making it difficult for clinicians and patients to trust the system's decisions.

Finally, we discuss the practical limitations of utilizing robotic assessments. The high cost of robotic assessments is a significant barrier to the widespread adoption of these systems in resource-constrained clinical setups. The expenses involved in developing, deploying, and maintaining these systems include advanced hardware, sophisticated software, and regular technical support. The sheer cost disparity between traditional and robotic assessments often deters hospitals from investing in robotics. Furthermore, adopting a robotic system in neurodegenerative care requires patient cooperation, which may be very challenging for several reasons. Older adults, who represent a majority proportion of individuals affected by neurodegenerative disorders, may feel intimidated by or distrustful of advanced technologies. Additionally, patients with severe motor and cognitive impairments may struggle to interact effectively with robotic systems. Another factor to consider is the maintenance and reliability of the system. The robotic systems require regular maintenance, software updates, and troubleshooting to function optimally, necessitating a potential in-house support infrastructure. The need for specialized personnel to operate or troubleshoot these systems adds another layer of complexity and cost, which may discourage potential customers from widespread adoption.

9. Conclusion

In the last decade, robotics has emerged as a transformative tool in the assessment of neurodegenerative disorders by offering precise, data-driven insights into motor, sensory, and cognitive impairments. While traditional assessments still remain valuable, they are often subjective, resource-intensive, and lack the ability to detect subtle or minor changes over time. Robotic systems address these limitations by providing a more repeatable, objective measurement that enables continuous monitoring of patients that could enhance our understanding of disease progression in the patient. This, in turn, could greatly improve the quality of care offered to the patients. However, there are several challenges hindering the widespread adoption of robotics in both clinical and home settings, such as technological limitations, high costs, integration hurdles, limited patient acceptance, and ethical concerns over data privacy and security. It is vital to address these barriers which may require interdisciplinary collaboration and the development of more cost-effective and scalable solutions. Looking ahead, the future of robotic assessment should focus on the design and development of adaptive and multimodal systems that can be seamlessly integrated into clinical workflows and patient lives without much disruption. Furthermore, the advancements in AI, sensor technologies, telemedicine, and robotic systems can play a pivotal role in improving early diagnosis, personalized interventions, and the overall quality of care. Finally, as the field evolves, robotics may not only revolutionize how we assess individual patients and their complex conditions but may also contribute to a deeper and better understanding of their underlying pathology, paving the way for more effective treatments and improved patient outcomes.

Conflict of interest

The authors declare no conflict of interest.

Author details

Krishnasamy Tamilselvam Yokhesh
Independent Researcher, San Francisco, USA

*Address all correspondence to: yokhesh13@gmail.com

References

[1] Zahra W et al. The global economic impact of neurodegenerative diseases: Opportunities and challenges. In: Bioeconomy for Sustainable Development. Singapore: Springer Singapore; 2020. pp. 333-345. DOI: 10.1007/978-981-13-9431-7_17

[2] Olanow CW, Tatton WG. Etiology and pathogenesis of Parkinson's disease. Annual Review of Neuroscience. 1999; **22**(1):123-144. DOI: 10.1146/annurev.neuro.22.1.123

[3] Sheppard O, Coleman M. Alzheimer's disease: Etiology, neuropathology and pathogenesis. In: Alzheimer's Disease: Drug Discovery. Brisbane, AU: Exon Publications; 2020. pp. 1-22. DOI: 10.36255/exonpublications.alzheimersdisease.2020.ch1

[4] Wankhede NL et al. Etiology and treatment challenges for neurodegenerative disorders. In: Targeted Therapy for the Central Nervous System. Cambridge, MA: Elsevier; 2025. pp. 1-25. DOI: 10.1016/B978-0-443-23841-3.00001-7

[5] Hou Y et al. Ageing as a risk factor for neurodegenerative disease. Nature Reviews. Neurology. 2019;**15**(10): 565-581. DOI: 10.1038/s41582-019-0244-7. Available from: https://www.ipalrobot.com/

[6] Goetz CG et al. Movement Disorder Society-sponsored revision of the unified Parkinson's disease rating scale (MDS-UPDRS): Scale presentation and clinimetric testing results. Movement Disorders. 2008;**23**(15):2129-2170. DOI: 10.1002/mds.22340. Available from: https://www.ipalrobot.com/

[7] Tombaugh TN, McIntyre NJ. The mini-mental state examination: A comprehensive review. Journal of the American Geriatrics Society. 1992;**40**(9): 922-935. DOI: 10.1111/j.1532-5415.1992.tb01992.x

[8] Knudson M, Thomsen TH, Kjaer TW. Comparing objective and subjective measures of Parkinson's disease using the Parkinson's KinetiGraph. Frontiers in Neurology. 2020;**11**:570833. DOI: 10.3389/fneur.2020.570833

[9] Goetz CG, Leurgans S, Raman R. Placebo-associated improvements in motor function: Comparison of subjective and objective sections of the UPDRS in early Parkinson's disease. Movement Disorders. 2002;**17**(2): 283-288. DOI: 10.1002/mds.10024

[10] How Much Does an MRI Cost? Available from: https://www.compare.com/health-insurance/how-much-does-an-mri-cost

[11] Glover L. Available from: https://money.com/why-does-mri-cost-so-much/

[12] The Imaging Solutions Company. Available from: https://mxrimaging.com/Blogs/CT-Scanner-Cost-and-Benefits

[13] Radhakrishnan DM, Goyal V. Parkinson's Disease: A Review. Mumbai, India: Medknow Publications; 2018. DOI: 10.4103/0028-3886.226451

[14] Moustafa AA et al. Motor Symptoms in Parkinson's Disease: A Unified Framework. Amsterdam, Netherlands: Elsevier; 2016. DOI: 10.1016/j.neubiorev.2016.07.010

[15] Ballard C, Gauthier S, Corbett A, Brayne C, Aarsland D, Jones E. Alzheimer's disease. The Lancet. 2011; **377**(9770):1019-1031. DOI: 10.1016/S0140-6736(10)61349-9

[16] Barnes DE, Yaffe K. The projected effect of risk factor reduction on Alzheimer's disease prevalence. Lancet Neurology. 2011;**10**(9):819-828. DOI: 10.1016/S1474-4422(11)70072-2

[17] Cummings JL, Vinters HV, Cole GM, Khachaturian ZS. Alzheimer's disease. Neurology. 1998;**51**(1 Suppl. 1):S2-S17. DOI: 10.1212/WNL.51.1_Suppl_1.S2

[18] Bates GP et al. Huntington disease. Nature Reviews. Disease Primers. 2015; **1**(1):15005. DOI: 10.1038/nrdp.2015.5

[19] Kiernan MC et al. Amyotrophic lateral sclerosis. The Lancet. 2011; **377**(9769):942-955. DOI: 10.1016/S0140-6736(10)61156-7

[20] Zarei S et al. A comprehensive review of amyotrophic lateral sclerosis. Surgical Neurology International. 2015;**6**: 171. DOI: 10.4103/2152-7806.169561

[21] Crowe J et al. PID Control. London: Springer-Verlag; 2005. DOI: 10.1007/1-84628-148-2

[22] Rawlings JB. Tutorial overview of model predictive control. IEEE Control Systems Magazine. Jun 2000;**20**(3)38-52. DOI: 10.1109/37.845037

[23] Kumar SA, Chand R, Chand RP, Sharma B. Linear manipulator: Motion control of an n-link robotic arm mounted on a mobile slider. Heliyon. 2023;**9**(1): e12867. DOI: 10.1016/j.heliyon.2023.e12867

[24] Tamilselvam YK, Ganguly J, Patel RV, Jog M. Musculoskeletal model to predict muscle activity during upper limb movement. IEEE Access. 2021;**9**: 111472-111485. DOI: 10.1109/ACCESS.2021.3103438

[25] MIT Media Lab. Available from: https://www.media.mit.edu/groups/biomechatronics/updates

[26] Berkeley Lower Extremity Exoskeleton. Available from: https://bleex.me.berkeley.edu/wp-content/uploads/2021/04/S1-BLEEX-scaled.jpg

[27] Vanderbilt University Medical Center. Available from: https://news.vumc.org/2013/11/07/robotic-legs-healthy-limbs/

[28] Zeilig G, Weingarden H, Zwecker M, Dudkiewicz I, Bloch A, Esquenazi A. Safety and tolerance of the ReWalk™ exoskeleton suit for ambulation by people with complete spinal cord injury: A pilot study. The Journal of Spinal Cord Medicine. 2012; **35**(2):96-101. DOI: 10.1179/2045772312Y.0000000003

[29] Esquenazi A, Talaty M, Packel A, Saulino M. The ReWalk powered exoskeleton to restore ambulatory function to individuals with thoracic-level motor-complete spinal cord injury. American Journal of Physical Medicine & Rehabilitation. 2012;**91**(11):911-921. DOI: 10.1097/PHM.0b013e318269d9a3

[30] Romanato M et al. Quantitative assessment of training effects using EksoGT® exoskeleton in Parkinson's disease patients: A randomized single blind clinical trial. Contemporary Clinical Trials Communications. 2022;**28**: 100926. DOI: 10.1016/j.conctc.2022.100926

[31] Pan Y-T, Lamb Z, Macievich J, Strausser KA. A vibrotactile feedback device for balance rehabilitation in the EksoGT™ robotic exoskeleton. In: 2018 7th IEEE International Conference on Biomedical Robotics and Biomechatronics (Biorob). Enschede, Netherlands: IEEE; 2018. pp. 569-576. DOI: 10.1109/BIOROB.2018.8487677

[32] Palacios RH. Robotic Arm Manipulation Laboratory with a Six

Degree of Freedom JACO Arm. Monterey, California: Naval Postgraduate School; 2015

[33] Zhang Y, Chen Y, Zhuang M, Cao Z, Liu J, Tian J. Wearable Robots Improve Upper Limb Function in Stroke Patients. Singapore: Springer; 2023. pp. 534-542. DOI: 10.1007/978-981-99-6489-5_43

[34] Mang CS et al. Test-retest reliability of the KINARM end-point robot for assessment of sensory, motor and neurocognitive function in young adult athletes. PLoS One. 2018;**19**(1):94. DOI: 10.1371/journal.pone.0196205

[35] Firwana YMS et al. Regional cerebral blood perfusion changes in chronic stroke survivors as potential brain correlates of the functional outcome following gamified home-based rehabilitation (IntelliRehab)—A pilot study. Journal of Neuroengineering and Rehabilitation. 2022;**19**(1):94. DOI: 10.1186/s12984-022-01072-w

[36] Carroll K, Kennedy RA, Koutoulas V, Bui M, Kraan CM. Validation of shoe-worn Gait Up Physilog®5 wearable inertial sensors in adolescents. Gait & Posture. 2022;**91**: 19-25. DOI: 10.1016/j.gaitpost.2021.09.203

[37] Hoppe-Ludwig S et al. Usability, functionality, and efficacy of a custom myoelectric elbow-wrist-hand orthosis to assist elbow function in individuals with stroke. Journal of Rehabilitation and Assistive Technologies Engineering. 2021;**8**:20556683211035057. DOI: 10.1177/20556683211035057

[38] Goffredo M et al. Kinematic parameters for tracking patient progress during upper limb robot-assisted rehabilitation: An observational study on subacute stroke subjects. Applied Bionics and Biomechanics. 2019;**2019**:1-12. DOI: 10.1155/2019/4251089

[39] MIT Biomechatronics. Available from: https://www.media.mit.edu/groups/biomechatronics/overview/

[40] Berkeley Robotics and Human Engineering Lab. Available from: https://bleex.me.berkeley.edu/

[41] Teater RH, Zelik KE, McDonald KA. Biomechanical effects of adding an articulating toe joint to a passive foot prosthesis for incline and decline walking. PLoS One. 2024;**19**(5): e0295465. DOI: 10.1371/journal.pone.0295465

[42] Ziemnicki DM et al. Development and preliminary evaluation of a bimodal foot prosthesis for walking and running. Journal of Biomechanical Engineering. 2024;**146**(9):091010. DOI: 10.1115/1.4065516

[43] BrainGate. Available from: https://carney.brown.edu/research-projects/braingate

[44] Deo DR, Willett FR, Avansino DT, Hochberg LR, Henderson JM, Shenoy KV. Brain control of bimanual movement enabled by recurrent neural networks. Scientific Reports. 2024;**14**(1): 1598. DOI: 10.1038/s41598-024-51617-3

[45] Collinger JL et al. 7 Degree-of-freedom neuroprosthetic control by an individual with tetraplegia. Lancet. 2013; **381**(9866):557-564

[46] Rewalk. Available from: https://exoskeletonreport.com/product/rewalk/

[47] Ekso Bionics. Available from: https://eksobionics.com/ekso-indego-personal/

[48] Marchese R, Diverio M, Zucchi F, Lentino C, Abbruzzese G. The role of

sensory cues in the rehabilitation of parkinsonian patients: A comparison of two physical therapy protocols. Movement Disorders. 2000;**15**(5): 879-883. DOI: 10.1002/1531-8257 (200009)15:5<879::AID-MDS1018>3.0.CO;2-9

[49] Tamilselvam YK, Jog M, Patel RV. Robot-assisted investigation of sensorimotor control in Parkinson's disease. Scientific Reports. 2023;**13**(1): 4751. DOI: 10.1038/s41598-023-31299-z

[50] Tamilselvam YK, Jog MS, Patel RV. Robotics-based characterization of sensorimotor integration in Parkinson's disease and the effect of medication. IEEE Transactions on Neural Systems and Rehabilitation Engineering. 2023;**31**: 3201-3211. DOI: 10.1109/TNSRE.2023.3299884

[51] Metta G et al. The iCub humanoid robot: An open-systems platform for research in cognitive development. Neural Networks. 2010;**23**(8–9): 1125-1134. DOI: 10.1016/j.neunet.2010.08.010

[52] Wood LJ, Zaraki A, Robins B, Dautenhahn K. Developing KASPAR: A humanoid robot for children with autism. International Journal of Social Robotics. 2021;**13**(3):491-508. DOI: 10.1007/s12369-019-00563-6

[53] Shamsuddin S et al. Humanoid robot NAO: Review of control and motion exploration. In: 2011 IEEE International Conference on Control System, Computing and Engineering. Penang, Malaysia: IEEE; 2011. pp. 511-516. DOI: 10.1109/ICCSCE.2011.6190579

[54] Robaczewski A, Bouchard J, Bouchard K, Gaboury S. Socially assistive robots: The specific case of the NAO. International Journal of Social Robotics. 2021;**13**(4):795-831. DOI: 10.1007/s12369-020-00664-7

[55] Kang HS, Makimoto K, Konno R, Koh IS. Review of outcome measures in PARO robot intervention studies for dementia care. Geriatric Nursing (Minneap). 2020;**41**(3):207-214. DOI: 10.1016/j.gerinurse.2019.09.003

[56] iPAL Robot. Available from: https://www.ipalrobot.com/

[57] Boyer M, Bouyer L, Roy J-S, Campeau-Lecours A. Reducing noise, artifacts and interference in single-channel EMG signals: A review. Sensors (Basel). 2023;**23**(6):2927. DOI: 10.3390/s23062927

[58] Abbruzzese G, Marchese R, Avanzino L, Pelosin E. Rehabilitation for Parkinson's disease: Current outlook and future challenges. Parkinsonism & Related Disorders. 2016;**22**(Suppl. 1): S60-S64. DOI: 10.1016/j.parkreldis.2015.09.005

[59] Lee TK, Yankee EL. A review on Parkinson's disease treatment. Neuroimmunology and Neuroinflammation. 2022;**8**:222. DOI: 10.20517/2347-8659.2020.58

[60] Tamilselvam YK, Ganguly J, Jog MS, Patel RV. Sensorimotor integration: A review of neural and computational models and the impact of Parkinson's disease. IEEE Transactions on Cognitive and Developmental Systems. 2024;**17**(1): 1-19. DOI: 10.1109/TCDS.2024.3520976